PEDESTRIAN PRECINCTS IN BRITAIN

John Ro

ABSTRACT

In Britain, shopping streets primarily for pedestrians have a venerable history. The earliest in this book dates from the Thirteenth Century. Many Victorian and later arcades are still extensively used, and the early post-1945 New Towns carried on the tradition of providing some traffic-free shopping streets. However, in the conversion of traditional shopping streets to pedestrian precincts, and in the construction of enclosed shopping centres, Britain was a late starter. West Germany's first conversion was in 1929, the USA's in 1959, and Britain's in 1967. Since then growth has been very rapid, so that there are few towns or cities now without a shopping precinct.

This book investigates the number, type and location of these precincts within the UK. It then considers their dimensions and the kinds of face-lift that most have received, providing a pleasant and safe environment in which to walk and shop. A Chapter looks at the way that various types of transport convey people and freight to and from the precinct. When people have reached it, further Chapters consider how they behave, what attitudes they have to it, and how they spend their money.

The main text concludes with a look at what the future might hold, and comments on the marvellous yet minute contribution that precincts make to the needs of pedestrians in towns. A large tabulation of 1304 known pedestrian streets, arcades or enclosed centres is followed by an 180 item bibliography.

> *COVER*
>
> *The five cover diagrams show the growth of Chester's pedestrian network from the Thirteenth Century and later Middle Ages to 1981, with intermediate additions at 1966 (including a 1910 arcade), 1972 and 1973. The text refers to these diagrams in para. 2.14.*

TRANSPORT & ENVIRONMENT STUDIES

supported by a grant from the Rees Jeffreys Road Fund

Published by
Transport & Environment Studies (TEST)
103-107 Waterloo Road London SE1 8UL
England
Telephone 01-633 0033
Cables TEST LONDON SE1

Printed and bound by High Speed Printing Limited

ISBN 0 905545 02 8
Copyright © TEST 1981

In memory of Alice May Whipp

CONTENTS

iv

ILLUSTRATIONS

PREFACE

This book has two origins. The first was a realisation that Monheim's work on German precincts, and Brambilla et al's on North American precincts, had no parallel in Britain. While Britain was comparatively late in a large-scale introduction of precincts, by 1980 it had well over 1000 streets, arcades and enclosed centres where people could shop in traffic-free or radically-reduced-traffic circumstances.

Coincidentally with this desire to fill a gap in the literature, Transport & Environment Studies (abbreviated to TEST from this point onwards) was asked by London Transport in 1980 to prepare a literature/state-of-the-art review of 'Buses and Pedestrian Areas' which they published a few weeks before this book. Obtaining the data for the London Transport publication was a daunting task. Suitable material was not held by the Department of Transport, so an elaborate postal survey was set up, initially of all Chief Planning Officers or their equivalent in the United Kingdom; this was followed by letters to certain District Councils, and reminder letters. The wealth of information received was far more than was required for the London Transport publication. We therefore asked them if we could use the excess material as the foundation for this book: they readily agreed.

Three problems followed. First, how to finance the project. Second, how to improve the data base. Third, how to find the time to analyse all the material (for it seemed unsuited to computer analysis) and write this book. The first problem was partly solved by a generous grant from the Rees Jeffreys Road Fund. The Trustees of the Fund made no stipulations about the content of the book, though they agreed its broad synopsis; doubtless they would wish to say that they do not necessarily agree with the views presented here. The relatively high cost of the book nevertheless is a reflection of the demands placed on a small applied research organisation.

In order to solve the second problem, improving the data base, a further round of letters was sent out to a large number of District Councils asking for informational gaps to be filled. At the same time literature searches were carried out, aided by computerised bibliographies prepared by the Transport & Road Research Laboratory and the Greater London Council for the London Transport project. This is an appropriate point to express our very sincere thanks to the literally hundreds of people involved in provision of all this information.

Solving the third problem involved people closer to home, within TEST's staff and associates. Mary Baker and Teresa Ryszkowski undertook literature searches. The data were analysed mainly by Terry Woolmer.

*John Elkington critically reviewed the book and made many valuable
suggestions about errors of fact, style or polemics. Ann Whipp also
read the text and added her cogent comments. Jenny Maynard and John
Roberts shared the typing, and also prepared the graphics with Terry
Woolmer. John Roberts took the photographs of precincts, Norman Elliott
those of the type face for cover and chapter headings. The book was
assembled at various times between June 1980 and March 1981.*

*Some explanations are needed. First, we have tried to make the book
readable, with a pyramidal structure. At the apex of the pyramid is
the Abstract on the title page. Chapter 0 summarises the book. Each
subsequent Chapter has a short resume. Second, all tables, graphs,
histograms and photographs have a common name: they are all termed
Illustrations. All the Illustrations have Roman numerals equivalent
to the Chapter number, followed by consecutive Arabic numerals, all
in Italic (eg IV.07). This system differentiates from paragraph numbers, which are all Arabic, and in Letter Gothic lettering (eg 2.15).
Third, metric dimensions are used normally unless imperial measures
are being quoted: in the latter case, they are also converted to metric
units.*

*The tabulation in Chapter 10 transmits what we have been able to col-
lect about precincts UK-wide; we wish we had had the resources to
provide a plan of each precinct, but this would have meant a document
of about 1000 pages, and prohibitive cost. The tabulation undoubtedly
contains errors and is incomplete by, we estimate, about 150 units.
We should be most grateful for the errors to be corrected and for
additional information to be provided (in the level of detail, and
using the symbols, of the tabulation please) which could be incorpor-
ated in a second edition. Any other comments on the book would be welcomed.*

*This text simplifies the mass of primary data held. Those wishing to
pursue the subject in more detail, or seeking the full evidence for
a particular presentation, may wish to buy the statistical supplement
to this volume. As its standards of design will be lower than for
this book, its price will be correspondingly less. Please write for
details. (Should you wish to buy the London Transport publication on
'Buses and Pedestrian Areas', at £3.50 plus postage, please apply
direct to them at 55 Broadway London SW1).*

*John Roberts
London, March 1981*

viii

0
summary

0.01
This book is about walking. As this is a very large subject, the main discussion is about shopping areas that have been set aside for people walking. Larger areas are considered only to provide a context for such developments, and to consider the choice shoppers make about the type of transport they will use to reach the shops.

0.02
We first make some comparisons between the three types of transport most used for relatively short distances - walking, private car, and bus (some longer excursions are made to large suburban or out-of-town shopping centres, but the importance of local and town centre shopping is not excessively diminished as a result.) Walking, or car, or bus, may be the most used form of transport in any particular place, and the wide differences in use are explained by level of car ownership, quality of the bus service, availability of car parking, ease, safety and pleasantness of a walk to the shops, and so forth. In terms of their impact upon other life and, by inference, the wider environment, people do little damage when they walk and a great deal when they travel by car. Buses, among motor vehicles, create the least impact when judged on a people carried x distance basis. The provision of pedestrian precincts is therefore strongly welcomed. The low level of facilities for pedestrians elsewhere in the town is deplored, however, for this is completely out of phase with the importance of walking to the shops, for which much evidence is provided.

0.03
The creation of pedestrian precincts is not just a matter of providing safety for people walking: there are many other reasons. Where it is done in town centres this may be to counteract the flow to suburban, out of town, or mail order shopping. It may have been civic pride, or have been connected with urban renewal, or have been part of the conservation of an historic centre. Whatever the reason, the conversion of all-traffic streets to ones primarily for people walking has progressed at an accelerating pace since the first was introduced in Essen, Germany, in 1929. The first of this type in the USA was in Kalamazoo, Michigan in 1959, while the probable first in Britain was London Street, Norwich in 1967. Britain lacked the legislation for conversions until 1967, but was well up in the world league pre-1939 with its many covered arcades, and in newly built schemes post-1945 in its New Towns. The other type not so far mentioned is the enclosed shopping centre, with its network of malls, of which there are now a large number in the UK.

0.04
Taking either a single converted shopping street, or an arcade, or an

enclosed centre as a 'unit' then this study, with the help of a large
number of local authority officers, has identified 1304 (90%) in the UK,
from its estimate that in 1981 there might be 1450 in all. Of the
1304, 119 (9.1%) are pre-1939, the remainder being post-1945 with
the 1970s as the peak time for their introduction (nearly 49% of the
1304). 91% of all identified units are in England. One third of all
the units are streets converted to pedestrian use, and 16.5% are newly
built, but uncovered, streets. The other sizeable type is the enclosed shopping centre, with 13% of the total number of units. Nearly
9% of all units share their movement space between buses and pedestrians.

0.05
No wholly satisfactory measure of individual counties' performance in
providing precincts has been found. In terms of the total number of
units, Greater London has the most, followed by West Midlands, Greater
Manchester, Cheshire, West Yorkshire, Lancashire and Durham. However,
63 of Greater London's 168 units date from pre-1939. English and Welsh
counties were then ranked according to population per unit (the first
three were Durham, Cheshire and North Yorkshire) and then on area per
unit (the first three using this measure were West Midlands, Greater
London and Tyne and Wear). Finally, all counties were grouped according to population density: in the Metropolitan/Highly Urbanised
group West Midlands performed best on an average ranking of population
and area per unit. In the Medium Urbanised group Cheshire came first;
in the Rural group, Durham, and in the Deep Rural group, North Yorkshire.

0.06
Having determined what types of scheme there are, where they are located, and which counties seem to be doing well in providing them, we
look at environmental factors: width and length of precinct, the
'cosmetic' treatment applied to them, and changes in noise and air
pollution figure strongly in Chapter 3. A rather inadequate sample
gives a mean width of street of nearly 20 metres, very similar to the
North American mean of 20.75 m. The length of 825 UK units was found
to average 150.7 m, only about one third of the US mean of 443 metres.

0.07
A study in Central Birmingham showed what people most preferred about
precincts. The three first preferences were 'things to look at such
as shops, posters, kiosks etc', 'freedom from crowding on footpaths',
and 'freedom from traffic.' As might be expected there is a wealth
of evidence that noise and air pollution levels have reduced quite noti-
ceably since traffic was removed from shopping streets.

0.08
In Chapter 4 we return to a more detailed look at how all manner of
people and goods reach and leave pedestrian precincts. The movement
of people is undertaken in more ways than the three principal means
suggested earlier: not only should bicycles, powered two wheel vehicles, light vans and rail be considered, but the needs of disabled
people and of emergency traffic are significant too. It is unfortun-
ate that the disabled person's 'Orange Badge' scheme seems to be abused

4

in several cities: some are considering attempts to exclude such vehicles from precincts. Some surprises arise when the proportion of North Americans not using cars to reach shopping malls is calculated. For buses, a large number of references support the importance of this means of transport; 59 places are analysed on the relation of their bus stops to precincts and a reasonably healthy result is derived: 313 bus stops (out of 740) were 50 metres or less from the precinct, the remainder being 51-200 metres distant. The importance of underground or Metro rail systems for shopping centres in large cities is underlined.

0.09
Delivery and collection of goods at shops has been hotly debated for many years. While USA and West Germany seem only moderately dependent on servicing other than at the shop frontage (and that normally occurs outside the peak periods of pedestrian use) the UK seems to have become rather obsessed by having rear, side, under or above, servicing, seemingly for two reasons: because the goods carriers want to use large vehicles, and because the precincts' designers (and users) want to keep pedestrian ways free of vehicles wherever possible. Of course it is sensible to service newly constructed precincts from the rear, above or below, but nevertheless, looking at 800 post-1945 UK units identified in this study, 39% are serviced from the pedestrian street or by trolley, nearly 50% are serviced from any other location than the frontage, and for the remainder, some combination applies. A wide variety of time constraints is applied, and violations occur, often because of difficulties besetting the supplier.

0.10
It is very encouraging that accident levels within precincts have dropped substantially from their all-traffic days; there is also, in general, an excellent safety record in those streets shared by buses and pedestrians. Few will be surprised about the reduced accident rate, or of the fact that the number of pedestrians in a precinct generally increases greatly over the previous condition. Crowding has also lessened: this was not only considered a nuisance by many (as we have already noted for Birmingham), it was also dangerous if you were forced from pavement to carriageway, there to mingle with motor traffic.

0.11
Most people have a viewpoint about pedestrianisation and often that view differs according to the interest group to which they belong. It is safe to say that among shoppers there is a high level of acclaim for traffic-free shopping areas, in virtually all parts of the country. Opinions differ about whether buses should share movement space with people walking, and many people complain about the number of parking spaces or their distance from the precinct. Sometimes surveys produce a mandate for pedestrianisation (as in Hampstead, London) which seems to be ignored, sometimes they influence all aspects of schemes.

0.12
Shopkeepers have strong views about pedestrianisation because it can

improve or diminish their turnover. For many years there was almost
violent reaction against pedestrianisation, though more recently this
has lessened considerably, probably because it can be seen how successful most precincts have been. It may be that a shopkeeper in an un-
pedestrianised street will suffer, as might one where the flow of people
past his or her door has reduced since a precinct was created. However,
in 15 out of the 18 cities on which data were collected by the Civic
Trust, trade had increased. OECD studied 105 cities worldwide and found
that 49% of the pedestrian zones recorded upward trends in turnover
rates. Sometimes, as with London's South Molton Street, pedestrianisation can be too successful: the traders there experienced demands for
excessive increases in the rents for their shops.

0.13
Looking to the future it may be that the peak of pedestrianisation has
been passed, though a great deal remains to be done. Many towns still
have no pedestrianisation (sometimes they are ones where it was vetoed
by the traders) and London's poor performance in recent provision of
precincts certainly needs remedying. A number of major shopping streets
may not have been pedestrianised simply because of the difficulties of
dealing with displaced traffic. A suggestion is made that such streets
should be looked at anew, and their total movement space re-allocated
between users on criteria such as number of persons moved per metre per
second, to use a scientific approach, or based on more abstract concepts
like serendipity or fun if your concern is social. Publications about
planning or transport futures for particular areas are usually disap-
pointingly bereft of positive thinking about pedestrians. Lateral
thinking rarely gets further than Edward de Bono.

0.14
In conclusion, 1304 units, or the total estimate of 1450 in the UK, are
in most cases a delight for their users: they are very much to be encouraged. However, and accepting that there are other traffic-free areas
around housing, educational, sports and medical campuses, and elsewhere...
then, if we compare the total GB public road network with the GB units'
total length, we find a ratio of 1563:1. In Greater London, for purely
urban roads, the ratio is 1132:1; in Doncaster it is 765:1. There seems
to be a major discrepancy, particularly in areas where a comparatively
low level of pedestrian activity incurs high accident rates because of
a false sense of security: ie, suburban areas.

0.15
Much of our urban environment remains an unsuitable place for its walking
residents and visitors. A suggestion is made that 'motorist precincts'
should be created above or below ground, or concentrated in corridors
well away from people behaving normally (ie walking), with thin cul-de-
sac filaments permitting access to those buildings that really have to
have vehicles alongside. The rest of urban movement space (at least
many roads) would then be converted to places which people could
really enjoy, and not only where they can move around in safety.
Towns must permit and encourage social communication, allow innova-
tory things to happen, and admit randomness. Much of these need to
take place out of doors. They are unlikely to do so as long as so
much of our urban space is devoted to motor traffic.

shopping in safety

As a generality, most people reach shopping centres either on foot,
by bus or by car. Walking there, when possible, makes very good
sense. Going by bus, if your town is lucky enough still to have a
quality service, is beneficial to society. Going by car is, in many
cases, unjustified and detrimental to society.

When you arrive at the shops you may well be in one of the 1450 or
so UK shopping 'units' - single streets, arcades or enclosed centres -
that have consciously been set aside for people walking. The decade
of the Seventies saw the greatest number arise in the UK, though most
are in England.

Those precincts were created partly to safeguard people moving on
foot, but there were many other reasons, some arbitrary. If the only
reason was the walker's safety, then streets for people would be found
city-wide. They are not. It is therefore surprising that the large
number who walk all the way to the shops arrive, mostly, still in one
piece.

Conservation of existing shopping streets (those that were converted
to precincts are one third of the total number of all types of shopping precinct) seems to have been strongly preferred to other, more
innovative, solutions.

The Chapter ends with a brief look at other works on the pedestrian.

I.01 : Pedestrian street, with butcher and baker, Robin Hood's Bay North Yorkshire

Walking: an almost free good

1.01
People are most likely to travel to a shopping area in one of three
ways: on foot, in a car, or by bus. Illustr. *I.04* shows this clearly. We need briefly to compare these forms of transport in order
to appreciate the importance of walking, which was not even considered a 'mode' of transport worthy of statistical analysis until the
mid-1970s, in Britain at least.

1.02
Starting with the private car, this has as many defenders as detractors, often the same people. A small proportion of the defenders are
essential users, while the detractors often have universal truths on
their side. Naturally this split of beliefs has generated a large
general literature: to sample this, see Papanek (1974) on obsoles-
cence; Thomas and Potter (1977) for predictions of the car's dis-
appearance; Plowden (1980) who believes it can be tamed; Bendixson
(1977) for alternatives; Potter (1981) on the billion pounds a year
tax evasion through company cars; Town & Country Planning (1980) on
the car's energy consumption when compared with public transport;
and Potter (1980) for some comments on the relationship between car
ownership and income: 95% of households he interviewed in Social
Classes I and II had a car, while only 43% of those in Social Classes
IV and V had one.

1.03
While the car is very useful in some situations, in others - and this
can mean many parts of cities and towns - its adversities outweigh
its benefits if society is viewed collectively and not just as an
aggregation of the satisfaction of individual wills and impulses.
Many have considered the costs of personal mobility. Few are prepared to do anything about them. Most people will retain their car
for a variety of reasons, sometimes practical, more often tenuous and
less supportable in an oversubscribed world. Take just one example
by Walter (1981): he suggests that much car travel is in the nature
of a gift from owner, or only-driver, to friends or family. In the
case of family it tends to be a patronising act between 'car owner'
(often still the senior male, who will talk about 'my' car) and his
wife or children.

1.04
On the other hand buses, by their nature, possess few of the unpleasant features of the car (though some would argue they have their own

9

disturbing characteristics), are disappearing because of the car, and in many places suffer negative discrimination. This form of transport that has the best ratio of people carried to movement space has to pull off the road at bus stops, then experiencing problems in rejoining the traffic stream. Its stops are often not where they are needed by passengers (one obvious location is at road inter-sections where they *are* placed in car-conscious New York). Often stops are remote from interchange with other transport; traffic management sends buses and passengers on unwanted tours of back streets; bus lanes are blocked with illegally parked cars, and so this listing of concerns could continue. Where buses *do* cause acci-dents, or pollute the air, these impacts are more thinly spread when related to the total person-km travelled in one bus.

1.05
For healthy people the only real criticisms of walking are that the distance you can reasonably travel is limited, and that it is a fair-weather and desirably a flat-terrain mode. It is reasonably effi-cient in energy terms but does consume footwear. (There is discus-sion on the costs of walking, and the costs of *not* walking, in Pushkarev and Zupan (1975). One important cost to the walker is the time taken for the journey). Walking has a venerable history and represents a largely free transport facility to the State. This has been little recognised, though we are told by the Department of Transport (1980) that two out of every five journeys are on foot. If walking is so significant, then it would be reasonable to expect it to be well-provided for: in some special places, the primary sub-ject of this book, this is so. Over cities and towns as a whole, it is not. Roberts (1980) highlights problems of transport self-suffi-ciency - the hazards associated with walking. As these will tend to recur through this book, they do not need detailed restatement here. How significant *is* walking to the shops?

Walking to the shops

1.06
The characteristics of walking in towns not unnaturally vary with the purpose of the journey. So, as Todd & Walker (1980) point out - Illustr. *I.02* - shopping is the most important of these purposes, although there will be considerable variation in its importance con-nected with the size of the area and distances between where people live and where they work, shop and are educated, etc. If we then look at numbers of roads crossed, and distances walked by day of week (from Todd & Walker and shown in Illustr. *I.03*) we can deduce that 'shopping' again outpaces the other purposes. The figures for shopping

in Illustr. *I.03* are shown with the next highest purpose in parentheses.

Purpose of journey	Number of roads crossed %	Time spent walking %	Distance walked %
Work or education	26	17	24
Shopping	37	47	36
Social or entertainment	23	19	23
Walks or trips	6	10	10
Other	8	7	7
	100	100	100

I.02: Distribution of activity measures by purpose of journey.

Shopping trip measures	Mon-Thurs	Fri	Sat
Mean No. of roads crossed per day	3.0 (3.0 work or education)	4.2 (2.8 work or education)	6.0 (2.6 social or entertainment)
Mean distance walked per day, km	0.62 (0.63 work or education)	0.87 (0.58 work or education)	1.35 (0.57 social or entertainment)

I.03: Roads crossed and distances walked for shopping purposes.

1.07

Over the whole week it is notable that the mean distance walked in km/day for men is 0.43 km and for women over twice that amount at 0.93 km. TEST's work in Brockley (TEST 1974), in Kentish Town and Putney (TEST 1976), and in Sutton (TEST 1980) shows mean walking distances as follows. In Brockley, LB*Lewisham, from 896 people interviewed the mean walking distance for shopping purposes on a Tuesday was 1180m, while on a Saturday it was 1135m. In Kentish Town, LB*Camden, it was 900m for Tuesday and Saturday combined, and the corresponding mean for Putney High Street, LB*Wandsworth, was 920m. Finally, in Sutton High Street for all purposes, Tuesday and Saturday combined, the mean walking distance was 980m. While people aged 60+ walk less than
(* = London Borough)

11

younger people, the differences are very moderate for shopping trips, and the elderly are much more at risk. (See Sheppard & Valentine 1980).
1.08
Two issues of'Social Trends' (CSO 1978, 1980) enable comparisons to be made between the National Travel Surveys in 1975-6 and 1978-9. The first table below (Illust. *I.04*) shows that there has been a moderate proportional increase in walking for shopping and personal business purposes while the use of private cars has moderately declined, along with bicycles and buses. (There is more detail on the means of reaching a shopping area in Chapter 4). The next table (Illust. *I.05*) shows scarcely any change in the proportional distribution of journey distances, by all forms of transport, for shopping and personal business purposes.

Means of transport used for major part of journey	% A=1975/6 B=1978/9					
	To and from work		Education		Shopping and personal business	
	A	B	A	B	A	B
Walk	18	20	60	61	43	46
Bicycle	5	4	4	3	3	2
Motorcycle	2	2	-	-	1	1
Bus	13	17	13	13	14	13
Car	55	-	13	-	39	-
car driver	-	34	-	2	-	20
car passenger	-	12	-	12	-	16
Rail	-	3	-	1	-	1
Other	7	7	10	7	2	2

I.04: Changes in means of transport for various purposes 1975/6 and 1978/9.

1.09
Both of these tables (*I.04 and I.05*) show a remarkable constancy over the three year period. The National Travel Survey of 1972/3 (DOE 1975) has different descriptions from the later Surveys: walking journeys of less than 1 mile are excluded. Thus there are no reliable comparisons over a reasonable time period such as 10 years and the cancellation by the Department of Transport of the 1981/2 National Travel Survey can hardly help those wishing to determine longer term trends.

	% of all shopping and personal business journeys	
	1975/6	1978/9
Under 1 mile	38	40
Over 1 and less than 3	33	32
" 3 " " " 10	23	23
" 10 " " " 15	3	3
" 15 " " " 30	2	2
" 30 " " " 100	1	1
Over 100	–	–

I.05:   Changes in distances travelled for shopping and personal business 1975/6 and 1978/9.

1.10
However, Hillman & Whalley's (1979) analyses of the earlier National Travel Surveys while not furthering long-term trend data, do show the proportion made on foot, by distance of all shopping journeys:

	% of all shopping journeys on foot	all methods	% of all shopping journeys in this distance band made on foot
< 1 mile	38	42	90
1 mile but < 2	10	22	45
2 " " " 3	2	11	15
3 " " " 5	< 0.5	11	2
5 or more	< 0.5	14	< 0.5
all distances	46	100	46

I.06:   Distribution of shopping journeys by distance and travel method.

Hillman & Whalley go on to note that walking is the method of travel of three-quarters of the journeys covering less than two miles - which in turn represents two-thirds of all shopping journeys. They, and others, devote more space to different behavioural patterns by age, sex, and access to a car than we do here, though one final comment from the 1975/6 National Travel Survey (Department of Transport 1979) might be made. The Survey's analysis of preferred means of transport for all journeys, by day of week, shows that walking is used for over 50% of journeys on Sunday to Wednesday inclusive. On Thursday it is 47%, and on Friday it is 46%, while on Saturday it drops to 35%, that being the day with the greatest car use - 45% as against a range of 26%

to 37% on other days of the week. This leaves us in no doubt about the importance of walking, though we should be aware that we are talking about journeys, not journeys x distance: most walks are short, as we have noted, while most car journeys are considerably longer (for a breakdown of unpublished National Travel Survey data on this topic, see Thomas and Potter (1977)). The lower proportion of walking on Saturdays may be reflected in a lower proportion walking to shops, and a higher proportion using cars for this purpose. If this is true, then the pedestrian will have even greater need for caution!

1.11
Paragraphs 1.06-1.10 strongly support the earlier contentions in this Chapter that people walking in towns need a safe and attractive environment. Most walking in towns is for educational purposes, primarily to and from schools. Shopping and personal business together form the next highest proportion of urban walking, as Illustr. *I.04* shows. Unlike work journeys, educational and shopping journeys are often quite short and it would seem that something could and should be done to make such journeys safe and attractive for people walking. However, while very little has been done to improve walking conditions on routes to a shopping area, quite a lot has been done to improve those conditions when major shopping areas are reached.

Why set aside shopping streets for pedestrians?

1.12
Why, then, have many shopping centres been pedestrianised, usually in part? Perhaps we should distinguish between streets that have been converted from all-traffic streets to those dominantly for the use of pedestrians, and the newer types: purpose-built streets and covered shopping centres. Arguably in the former case streets were converted because of civic pride, or to stem the outflow of trade to peripheral shopping centres (a report on pedestrianising The Commons, Ithaca, New York is actually sub-titled 'Bringing People Back to the City' (Van Cort 1978)); because city fathers had seen a scheme in say Germany and had returned to impose pedestrianisation on rather unwilling occupants and users of the street, or even because a new road around the centre had left the shopping streets relatively free of vehicles. It is relevant to note that pedestrianisation has rarely resulted from shopkeeper initiative - more often they reacted against the concept, later to be proved wrong (see Chapters 6 and 7).

Purpose-built shopping precincts, on the other hand, are naturally adapted to walking - all the other associated traffic can easily be placed out of sight.

14

1.13

There is not much information available on the reasons why shopping areas in Britain have been pedestrianised, though an analysis has been done for West German streets by Monheim (1975). Monheim established 12 reasons why a street should be converted for pedestrian use. He then invited upwards of 150 West German towns to apply a 6-point scale (1 = most important, 6 = least important) to six of these reasons. For this present book the frequencies with which the six points of the scale were applied were determined, and an approximate mean response was derived. These means were then ranked, with that one closest to 1 ranked first and so on, with the following result:

Monheim code	Reason for converting shopping street for pedestrian use (translated from the German)	Mean response on 6-pt scale	Total number of responses within scale
M	'Timely town planning, attractive image'	2.42	95
L	'Preservation of historical aspects of the town'	2.89	46
H	'Leisure value of the shopping centre, evening entertainment'	2.95	79
F	'Improved traffic circulation and safety'	3.11	88
B	'Attraction for shoppers from surrounding area'	3.49	78
G	'Less noise and air pollution'	3.49	76
C	'Less chance of losing shoppers to rival towns'	3.55	44
I	'To awaken civic pride'	3.62	13
E	'Less likelihood of losing shoppers to non-integrated shopping centres'	3.70	23
D	'Less likelihood of losing shoppers to suburban shopping centres'	4.00	5
A	'Attractiveness for visitors' (tourists?)	4.10	29
K	'Scope for political meetings, discussions and demonstrations'	5.40	5

I.07 : Ranked reasons for pedestrianising German shopping streets

1.14

Illustr. *I.07* reflects German cultural values, so some of the reasons, and their ranking, are less likely in other countries. Also codes I, D and K may have too few responses to be significant. Nevertheless, the most favoured German reason (which also had a high number of responses) is connected with urban renewal. Pedestrianisation may be a cause or an effect of this: contrary to expectation one British scheme occurred after construction of a central area by-pass, as something of an afterthought. In other localities, the desire to pedestrianise has *caused* by-passing; in yet others, various goals were pursued simultaneously. The second-ranked German reason is echoed in several other European countries (and in developing countries where their past is becoming valued); certainly in England there are many places where conservation of historic centres almost made pedestrianisation mandatory. The third-ranked, leisure-based, German reason may be less important in other countries, though Latin countries with obligatory promenading would doubtless support it. Most countries would agree with the safety aspects of the fourth-ranked reason. The National League of Cities (1980) has summarised the objectives concerning pedestrianisation of parts of a number of cities. Thus, for Curitiba, Brazil it was 'to reclaim urban space for pedestrian and cultural uses'; in Geneva the objective was rather more formal: 'To convert designated spaces to pedestrian areas'; in Osaka, Japan it was 'to protect the urban environment and improve the pedestrian environment'; and, for Oporto, Portugal it was 'to improve general road safety, especially for pedestrians.'

Growth in the number of pedestrian shopping precincts

1.15
Whatever the reasons for converting shopping streets for pedestrian use, there has been a dramatic increase in their introduction post-1945. Before this recent spate, reserving streets for pedestrians happened frequently in several countries over a long time period. Among the reasons then, it may have been desirable to have a formal processional way; a street may not have been wide enough for anything other than people walking; or, as with Venice, physical constraints kept out vehicles; or, climate meant shopping areas should be covered and heated or cooled (tropical cities like Singapore, cities and towns with severe winters like Montreal and northern Scandinavia). In the nineteenth century some European cities achieved covered arcades - the Galleria in Milan, arcaded streets in Turin, and many arcades in Britain are good examples.

1.16
In the twentieth century the first street conversion took place in Essen in 1929; the first in the USA was Kalamazoo in 1959; probably the first in Britain was London Street, Norwich in 1967. In some European cities - notably in the British New Towns - war damage rebuilding or new town construction encouraged purpose-built pedestrianisation. (Thomas and Potter (1977) however show how pedestrianisation was misconceived in the British New Towns, if the proportion of walkers to car users is a significant measure of the need to make provision for pedestrians. They found, quite simply, that people walk far less in New Towns than in historic towns: first, because ample parking at origin and destination and good roads in between encourage car use and, second, because land uses are segregated and often long distances apart - homes from place of work or shopping, for example).

16

More recently within the last twenty years, many city centres and more outlying areas have accepted - not always graciously or without damage to an historic environment - large new covered and often air-conditioned shopping centres.

1.17
An analysis of the cumulative rise in converted streets for USA, West Germany and UK was carried out for TEST's report on Buses and Pedestrian Areas (Transport & Environment Studies 1981). It was based on available data in mid-1980 and the graph is reproduced below as Illust. *I.08*. While the time periods and number of schemes are dissimilar for each country, each curve is similar to an exponential curve.

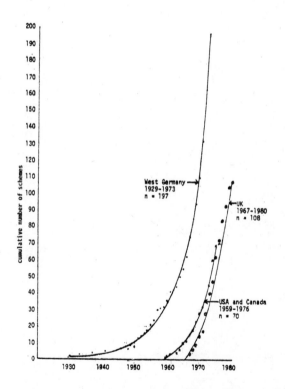

I.08: *Cumulative increase in streets converted primarily for pedestrians. UK, USA and Canada, and West Germany variously between 1929 and 1980. Note only those schemes for which data were achieved are shown.*

17

1.18
For this present publication TEST has obtained further information on British 'conversions' (ie. all purpose streets converted to those primarily intended for pedestrians), and 308 of these are portrayed in Illust. *I.09*. Here the curve is closer to a linear growth curve, though showing rapid if inconstant growth. 1967 was the year when the first modern legislation permitting conversions of highways was introduced (Road Traffic Regulation Act 1967). It was later supplemented by powers in the Town & Country Planning Act 1971. The powers in the two Acts are different: the 1967 Act is more appropriate for experimental closures, and allows fewer vehicle types to be excluded than the 1971 Act; however, the latter permits a greater range of type and works in the altered street. (Dept. of Transport 1978).

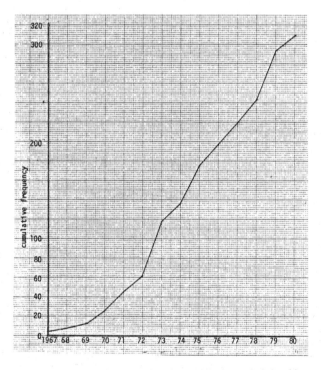

I.09: Cumulative growth in conversion of British all-purpose streets to ones primarily for pedestrians.

18

1.19

Illust. *I.10* derives from an OECD study of streets pedestrianised
within its member nations (OECD 1975). 444 cities were approached and
300 responded. The proportions with pedestrian zones are therefore
a little misleading as the number responding may be low. However,
the point of including this tabulation is to show how widespread is
the concept of pedestrianisation.

OECD Member country	No. of cities responding	% of cities with pedest-rian zones	considering ped. zones
German Federal Republic	30	87	37
Australia & NZ	13	62	38
Canada	17	24	29
USA	47	27	27
Spain	15	53	47
France	31	52	58
Italy	10	60	50
Japan	45	53	9
Norway	2	)	)
Denmark	4	)	)
Finland	1	)	)
Sweden	5	)	)
Ireland	2	)73	)27
Austria	4	)	)
Switzerland	5	)	)
Belgium	2	)	)
Netherlands	13	100	21
UK	50	76	12

*I.10: Proportion of OECD cities with or considering
pedestrian zones. Source: OECD 1975.*

Conservation versus technology

1.20

In the late 1960's and early 1970's there were several publications,
perhaps fired by Harold Wilson's 'White hot technology' speech of
the mid-sixties, in which this concept was applied to the
projected shopping of the future. Given the apparently substantial
financial resources of the time it is interesting that few of these
ideas got anywhere. Thus, Roberts (1970) contemplated what was then
called an Autotaxi being used not only for passenger traffic but

also to distribute goods purchased in a major shopping centre. That concept demanded first an Autotaxi network (Sheffield toyed with the idea but later abandoned it), second a great deal of new housing and services designed in conjunction with the network, and third, a population adjusted to phone ordering from catalogues and a Prestel-type TV with constantly updated consumer goods prices. In Czechoslovakia there were plans for pneumatic tube distribution of shopping purchases in the new town of Etarea. In the US there was excitement about credit cards being read by computer terminals at hypermarket checkouts - a concept just surfacing in Britain, ten years later. Certainly around 1970 there were many who believed that mail order, computerisation and novel distribution systems would rapidly revolutionise shopping in Britain.

1.21
The revolution fell flat on its face. What we were about to receive was enlargement of the same and conversion of the old type of street. Most first generation new towns adventurously built shopping streets with no vehicles and one or two enclosed shopping centres (as distinct from enclosed markets whose history stretched back to the nineteenth century or earlier) were built; in New Town terms Milton Keynes probably has recently upstaged all the others in the size and grandeur of its shopping centre (Illust. *I.13*). Reputedly this was designed with an 80km radius catchment area in mind, based on dramatically increasing car ownership; it is said that oil prices are restraining those who might have travelled long distances to it by car. Some hypermarkets peripheral to conurbations are apparently also suffering from the costs of reaching them. However, the trend to bigness is universal; the only problem for the consumer is that larger stores need larger catchment areas and therefore occur less frequently/conveniently/safely, for the user: a car is almost an imperative. That leads to large car parks surrounding the superstores - land made useless for other purposes.

1.22
In larger towns and virtually all cities one can find the frequently air-conditioned centre, based on two or three large stores connected by US-style malls, with smaller shops, restaurants, banks and children's play areas. There are notable examples in Birmingham, Newcastle, Leeds, Nottingham, Milton Keynes and many other places. Their rise in numbers is plotted on Illustr. *I.11*, which also contains a plot of the purpose-built but uncovered shopping streets, arcades and pedestrian streets. In Illustr. *I.12* the introduction of all types of unit known to TEST, by time, is displayed as a histogram: 1304 units have been identified. 119 (9.12%) date from before the second world war, but the decade after that war saw few additions to the stock. Enabling legislation in 1967 probably largely accounts for the additional 114 unts in the 1960s. Further legislation in the 1971 Town and Country Planning Act, and a general acceptance of the concept of pedestrianisation in Britain and worldwide, doubtless explain the addition of 636 units in the 1970s - nearly 50% of all the units achieved

by 1981. The other notable features of the histogram are the dominance of the B1 units (33%) and the rapid rise of A2 and C type units.

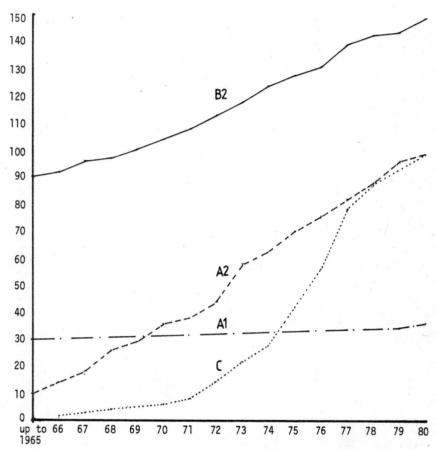

I.11: *Provision of four types of shopping street, cumulatively, by time*
A1 = covered street, eg. arcade, usually pre-1939
A2 = covered, purpose-built shopping centre, mainly post-1945
B2 = purpose-built, uncovered street
C = bus and pedestrian street

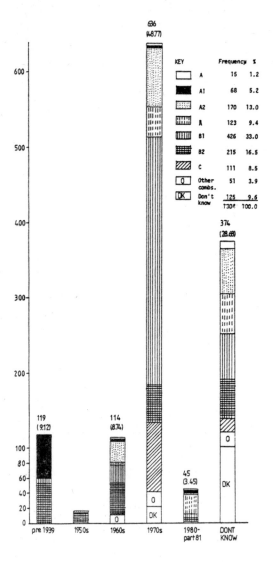

I.12 : *Introduction of all known units, by type and time*

22

1.23

The last group, already referred to in paragraph 1.17, is the street-converted-to-pedestrian-use. Streets of this type show practically no technological innovation. One feels they were converted either because, like Everest, they were there, or because they were historic streets that should no longer suffer the pounding of heavy traffic - or possibly both explanations hold. There does seem to be a very large number of these marriages - of commercial convenience and a planning system that favours conservation to anything too technologically advanced. The decision-makers and their advisers may well have right on their side: it is notable that TEST has identified 447 (34.3%) conversions of single streets of any date to early 1981 as against 214 (16.41%) purpose-built uncovered pedestrian streets, 64 (4.9%) arcades and 172 (13.2%) covered shopping centres. A further 117 (9%) streets permit buses to run along all or part of their length. The proportion in each case is of the total number of precincts (1304) - either single streets or a network of malls in a covered centre - identified in the UK. This total is larger than the total of the frequencies above as many streets are inadequately categorised in the information received.

1.24

Perhaps we should pick up some of this Chapter's early observations to end this part. Far from throwing out the car from a large proportion of the inner area of towns and cities, we appear to have limited this safety device to comparatively small sections where there happen to be shops or other central commercial or public buildings. Some towns still have no pedestrian streets and London's performance as a capital city is not exactly adventurous (but see paragraphs 8.03-8.05 below). The London Amenity and Transport Association (1973) made a plea for policy concerning pedestrians in London. Parker and Hoile (1975) showed just how many pedestrian streets there were in London, but did not underline the fact that most were pre-1939 and that most of those were nineteenth century creations. In 1981 some improvement has taken place but nowhere near consistent with the capital's size or prestige.

1.25

Reaching these pedestrian streets on foot can be quite hazardous and undignified for the sophisticated being we have ambiguously labelled 'pedestrian'. His or her conditions may only fully improve, ironically, as vehicle fuels become generally unaffordable or even unacceptable (it is now realised that the ultimate constraints will be first when economically recoverable fossil fuels run out, and second when growing food on the land is considered more important than growing vehicle fuels.) The time for these events will be outside the lives of most readers of this book.

Other works on the pedestrian

1.26
This Chapter has argued that the benefits of moving about in a car,
at least for many shopping purposes, are outweighed by its severe
impact on civilised life. This schism is much less noticeable in
public transport and scarcely at all in cycling and walking: walking,
despite its limitations, is presented as an almost free transport
facility to the State. The State, however, has shown some uncer-
tainty about accepting this gift. It was then suggested that at
least the journey to intensively used central area facilities like
entertainment, commerce and shopping would be biased toward the needs
of pedestrians. No such thing but, for a variety of reasons, the
pedestrian comes into his own at the actual point where these facilities occur. These precincts constitute a minute proportion of the
total area needed, city-wide, for people to walk pleasurably and
safely. This minute proportion has, however, received lavish attention worldwide with a rate of growth which at times has been exponential. To round off the discussion, some questions were asked
about why pedestrianisation as a shopping innovation was overwhelmingly preferred to other 'high tech' approaches that were feasible,
though not without problems. If the reader is concerned that this
argument and discussion are somehow incomplete, then some of the
many other writers' works are indicated in the next few paragraphs.

1.27
While Gray (1965) produced one of the earliest reviews of pedestrianised shopping streets in Europe, internationally the literature is dominated by OECD's many publications on pedestrians and on
traffic restraint policies that favour the more 'at risk' forms of
transport. Their 'Streets for People' (OECD 1974) remains one of the
better extant reviews, while a range of publications arose from the
'Better Towns with Less Traffic' Conference in 1975; the survey of
traffic limitation policies has been mentioned earlier (OECD 1975).
The Greater London Council (1972) reported on its study tour of
pedestrianised streets in Europe and the USA - a tour and report
that attracted much attention at the time, and activity within the
GLC, though this fairly soon gave way to a swing to the right in GLC
politics, and a seemingly parallel reduction in interest in pedestrianisation. Elkington et al (1976) sampled the world's literature
on the pedestrian and provided an extensively annotated 1000+ item
bibliography; at about the same time a shorter bibliography was
compiled by Copley and Maher (n.d.) A more recent 350-item bibliography is part of a brief work on planning for pedestrians (Ramsay
1980). Perkin (n.d.) compiled Europa Nostra's illustrated guide to
pedestrian precincts.

1.28
In the United States several important works on pedestrians have
appeared. Brambilla and Longo (1976 a,b,c,d) and Brambilla et al
(1977) are significant publications, though they were preceded by
three seminal works - Rudofsky (1969), who tried arousing Americans'

24

interest in pedestrian streets; Fruin's (1972) work on the pedestrian's right to urban space, which is one of many publications by this author in the 1970s; and (one of several of their works) Pushkarev and Zupan (1975), who were also concerned with design aspects and to whom we return in Chapter 5, paragraphs 5.07 and 5.08.

1.29
Two works from West Germans demand attention - Monheim's (1975) very thorough study of West German precincts, to which we shall return in the next Chapter; more recently, Boeminghaus (1979) studied pedestrian areas. Smith (1977) has looked at Mainz, West Germany, with British eyes.

1.30
In Britain, Alfred Wood fathered the new generation of shopping streets converted primarily for pedestrians, and he has described his Department's work while he was City Planning Officer in Norwich (Wood 1969). While there have not been many other practitioners who have published their work, Britain abounds in a theoretical, sometimes empirically-based, literature. Official Architecture and Planning (1970) devoted an issue to planning for pedestrians, while Brian Richards has several publications on or adjacent to the topic of pedestrian planning (see Richards 1976). Hillman and Whalley are also prolific writers, their concern with pedestrians often being with their 'second class citizen' status (see for example Hillman and Whalley (1979), already noted on page 13, and Hillman et al 1973 as examples: they organised a conference on walking in 1980 - Policy Studies Institute 1980). Like Hillman and his colleagues, Daor and Goodwin (1976) have noted how the frequency, purpose and length of walk vary for different socio-economic groups.

1.31
Having briefly reviewed the importance of walking, the benefits to be derived from encouraging it as a means of transport, and its significance as a way of reaching a shopping area, we can now move on to the different types of pedestrian shopping precinct, and where they are located.

(illustr. I.13 appears on page 26)

I.13 : Three views of Milton Keynes Shopping Centre

2

where and what?

This Chapter tells us about the various types of shopping street given over primarily to pedestrians, where they can be found, and the performance of different local authorities in making (or allowing developers to make) this provision.

Maps show the distribution of all known post-1945 units in the UK, and also the population per unit, post-1945, for England and Wales. Using these different measures provides different views of distribution. Pre-1939, 119 units were discovered, dominantly in England - 53% of them in London. However, there are known to be gaps in this collection.

Having looked at the various types (covered arcades or enclosed shopping centres, converted streets, and streets shared by buses and pedestrians are the main variants) the build-up of networks is considered, taking Chester, Coventry and Swindon as examples of how clustering of pedestrian streets occurs over time.

Finally some attempts are made to rank English and Welsh counties according to their provision, using several different measures.

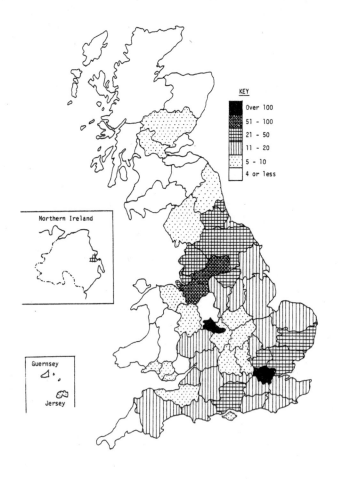

KEY

■	Over 100
	51 - 100
	21 - 50
	11 - 20
	5 - 10
	4 or less

Northern Ireland

Guernsey

Jersey

*II.01 : Frequency of units for United Kingdom, by County or equiva-
lent area, post-1945*

28

2.01
Two illustrations set the scene for this Chapter, showing the geographical distribution of units known to TEST and of all types - detailed comments on individual types and their date of introduction are provided later in the Chapter ('unit' is defined in the next paragraph). Illustr. *II.01* opposite shows the frequency of units by County or comparable area, from 1945 onwards. If pre-1939 schemes were included, then Greater London would have the highest number of units, followed by West Midlands, Greater Manchester, Cheshire, West Yorkshire, Lancashire and Durham. Over all time London would show 168 units, but 63 of these date from pre-1939. Those pre-1939 units on which information was received (notably excluding Leeds' nineteenth-century arcades) and using 1981 administrative areas are: Greater London 63; West Midlands 9; South Glamorgan 8; Greater Manchester 7; Avon 5; Dorset, Hertfordshire and Merseyside 3 each; East Sussex, Hampshire, Norfolk, Tyne and Wear, West Sussex, Highland, and Northern Ireland each 2; Cheshire, Durham, Clwyd, and Central Region each 1.

2.02
The map, Illustr. *II.02*, also displays (and ranks) only post-1945 units and then only in England and Wales. It is based on the 1971 Census adjusted for the new administrative areas post-1974 and depicts the population per unit for each County or equivalent area. This gives a different view from Illustr. *II.01*. The counties with the least population per unit are Durham, Cheshire, North Yorkshire (low population, large number of units in York), Isle of Wight, Suffolk, Lancashire, West Midlands. Greater London is ranked 49 using this crude measure. For more detail see paragraphs 2.17-2.19.

2.03
This Chapter deals with a geographical hierarchy of the following kind:

    i   United Kingdom
   ii   Major subdivisions (eg England, Wales)
  iii   Counties, Regions or equivalent
   iv   Districts
    v   Places
   vi   Precincts
  vii   Units

Definitions of the first four terms are self-explanatory for most people, but for completeness they are all defined below (the numbers are derived from Central Office of Information (1980)):

   i Literally United Kingdom of Great Britain and Northern Ireland.

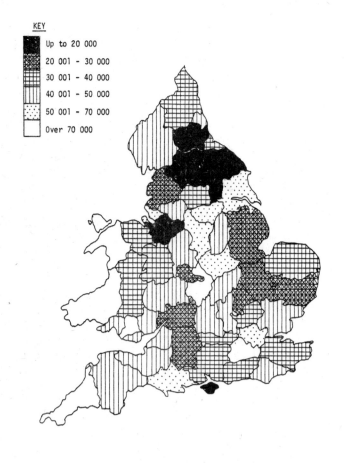

KEY

Up to 20 000
20 001 - 30 000
30 001 - 40 000
40 001 - 50 000
50 001 - 70 000
Over 70 000

II.02 : Population per unit, England and Wales, post-1945

	Great Britain contains England, Scotland, Wales and certain islands;
ii	The areas in i treated individually;
iii	Counties in England and Wales (Greater London, 6 Metropolitan Counties, 47 non-Metropolitan Counties), 9 Regions and 3 Island Authorities in Scotland, 26 Districts in Northern Ireland;
iv	District Councils are second tier administrative areas of England (36 Metropolitan, 296 Non-Metropolitan, 32 London Boroughs, City of London), Wales (37) and Scotland (53);
v	The term 'place' is used either for areas having a separate administration pre-local government reorganisation in 1974, or for those parts of urban areas that are distinctly recognisable locations. As an example, the West Midlands *County* has a *Metropolitan District* of Dudley within which Dudley, Halesowen and Stourbridge all have distinctly recognisable shopping centres and are therefore termed *places*;
vi	A precinct is any shopping area of any size which has been set aside primarily for pedestrians. It may be one street, or a number of connected units (see vii) when it may also be called a 'network'. The text distinguishes between 'single street precinct' and 'network precinct'; a place could have more than one precinct though this is infrequent;
vii	A 'unit' is one of the pedestrian facilities A1, A2, B1, B2 and C explained in paragraph 2.07 below. Thus it might be a single street, a single arcade, or a single covered shopping centre (even though it is realised that this often contains more than one mall: for all intents and purposes such a centre has a unity).

Elsewhere in the text other words are occasionally used for variety of expression. 'Facility' could be a precinct or a unit. 'Scheme' is usually used of one or more units established at much the same time - say within a five year period. 'Street' and 'mall' will be units. 'Arcade' is a covered street, and therefore a unit. 'Pedestrian Area' is clearly generic, being identical to 'precinct'.

2.04
The text contains frequent references to sub-samples of UK pedestrian areas. This is because no part of the information assembled for this book is complete, despite strong attempts to achieve a total set: it would be surprising if this had been achieved. The attempt started with a letter to the Chief Planning Officer of all UK Counties or equivalent level authorities. In some cases the letter was passed to another Chief Officer. The response was complex and is analysed in the statistical supplement to this book. A table in that supplement shows the broad response to these initial letters, and considers the second round of letters sent to certain District Councils, together with the response to the third round of letters requesting amplification of some initially received information. The few unhelpful authorities each received one or two reminder letters in addition.

2.05

The majority of Counties and Districts gave information unhesitating-
ly. Only one local authority, a Tory London Borough, made a charge
for its help, ironically providing information on only one unit, a
nineteenth century arcade. In order partially to fill the gaps in the
total set of information received, other sources were used, and the
final assembly of material may be found in Chapter 10, a substantial
part of this book.

Types of precinct and their distribution

2.06

A very broad division between pre-1939 and post-1945 units can be made,
by major geographical area:

Location	Pre-1939 Frequency	%	Post-1945 Frequency	%	All Frequency	%
England	105	88.2	1087	91.7	1192	91.4
Wales	9	7.6	41	3.5	50	3.8
Scotland	3	2.5	22	1.9	25	1.9
N Ireland	2	1.7	21	1.7	23	1.8
Channel Isl.			14	1.2	14	1.1
UK	119	100.0	1185	100.0	1304	100.0

*II.03 : Distribution of known units (see para 2.03) by time and major
geographical area*

The above table is confined to those units that were discovered during
the study. The total of 1304 units probably represents about 90% of
the total UK shopping streets (including enclosed centres) reserved
primarily for pedestrians: thus, there may be 1450 units in all at
March 1981. This figure was derived by thoroughly analysing all data
received from Counties, Districts or their equivalents, and comparing
this with the total of such authorities in the UK (using George Godwin
Ltd 1980): where, in the analysed data, there were no precincts in a
District this was explained by that simple fact, or by that District
not having responded to our requests for information. If it were the
latter explanation, the District's principal shopping towns were compared with others that were similar in size, location, population density etc., and an estimate made of the number of precincts. The 1450
are *shopping* units, however, and not the total of all pedestrian units.
That total would have to include the networks within educational and
medical complexes, housing developments, churchyards, and so forth.
Those wishing to know about that type of pedestrian facility, completed
under the 1967 or 1971 Acts (see paragraph 1.22) should refer to the
Department of Transport's lists (DTp 1980a).

2.07

In order to break down Illustr. *II.03*, the various types of unit need
to be classified. A relatively simple classification was evolved, and
used without question by all the local authority respondents. There
are three fundamental units, two of which are subdivided. Thus, type
A refers to covered shopping streets - arcades or malls (covered shopping malls are pre-eminent in the USA (see Bouchier 1981) and are rapidly assuming similar importance in the UK. A recent mammoth proposal
of this kind is for Whitechapel in London (Greater London Council 1980a).
With about 80 000 m^2 of shops this development would be larger than
Brent Cross in North London). Type B refers to uncovered streets that
have been primarily - though not wholly as with type A - given over to
pedestrian use. 'Primarily' has to be used, for virtually all of type
B allow access for emergency, and some for disabled persons', vehicles.
Some admit servicing vehicles, either all the time, or with restrictions
on time of day. Type C comprises uncovered units permitting buses to
share the movement space with pedestrians plus the occasional exceptions as for type B above. Finally, types A and B are further subdivided into A, A1 and A2, and B, B1 and B2.

In summary:

A   Covered units. Also used where subdivision A1 or A2 not
    known
A1 Almost entirely pre-1939, primarily arcades, but could
    include market halls, though these were excluded from this
    investigation (see Illustr. *II.04*)
A2 Purpose built, post-1945. Usually enclosed shopping centres
    with one or more 'malls' and one or more internal squares
    (*II.05*)
B   Uncovered streets. Also used when subdivision B1 or B2 is
    not known
B1 Originally all-purpose shopping streets containing all types
    of traffic, since converted to streets primarily for pedes-
    trians (*II.06*)
B2 Purpose-built, post-1945, uncovered. There are far less of
    these than of type B1: they occur mainly•in Mark I New
    Towns (*II.07*)
C   Type B streets that also permit buses (*II.08*).

Often, particularly with long streets, part is of one type, part of another; this happens fairly frequently,with type C occupying only one
end of a street which is otherwise B1.

Most pedestrian shopping streets are contained within the above categories. There are occasional exceptions like streets in Bracknell
and Farnborough (Betts 1975) which were newly built as all-purpose
streets and then later converted for pedestrians - in the case of
Queensmead Farnborough, after only 14 years.

33

*II.04 (top left) Hardye Arcade, Dorchester; II.05 (top right) Eldon
Square, Newcastle upon Tyne; II.06 (centre) Blandford Street,
Sunderland; II.07 (bottom left) Market Square, Corby; II.08 High
Street, Exeter*

2.08
The distribution of all these types, post 1945 and known from TEST's researches, is shown in the table below. The additional column 'combined' allows for streets with more than one category, while D/K is 'don't know'.

	A	A1	A2	B	B1	B2	C	Combined	D/K	All
England	14 1.18	5 0.42	158 13.33	77 6.5	399 33.67	155 13.08	108 9.11	46 3.79	125 10.6	1087 91.7
Wales			7 0.6	2 0.2	25 2.11	1 0.08	3 0.25	3 0.25		41 3.5
Scotland	1 0.08		6 0.5	3 0.25	5 0.42	4 0.34	2 0.17	1 0.08		22 1.9
N Ireland			1 0.08	15 1.26			4 0.34	1 0.08		21 1.8
Guernsey					3 0.25	1 0.08				4 0.34
Jersey				2 0.17	8 0.67					10 0.84
UK	15 1.26	5 0.42	172 14.51	99 8.35	440 37.13	161 13.6	117 9.87	51 4.3	125 10.55	1185 100.00
Greater London incl above in England			15 1.26		35 2.95	22 1.85	4 0.33	8 0.67	21 1.77	105 8.86

II.09 : Distribution of known shopping units by type and country post 1945

2.09
The lesser number of precincts constructed pre-1939 is tabulated below:

	A1	B1	B2	All
England	45 37.8	7 5.88	53 44.54	105 88.23
Wales	9 7.56			9 7.56
Scotland	3 2.52			3 2.52
N Ireland	2 1.68			2 1.68
UK	59 49.58	7 5.88	53 44.54	119 100.00
Greater London (incl. in England)	10 8.4		53 44.54	63 52.94

II.10 : Distribution of known shopping units by type and country prior to 1939

2.10

Illustr. *II.09–II.10* have several interesting features. First, there were few pedestrian precincts pre-1939, and these included nearly 50% of covered units (suggesting they were intended for the wealthy) - against less than 0.5% post-1945. Greater London has been included in the tables as it later merits special attention: it had 53% of all the know pre-1939 UK schemes but only about 9% of those created post-1945. However, London's exclusive possession of B2 units pre-1939 is suspicious - there must be other places in the UK that built new uncovered streets for pedestrians pre-1939; these may simply not have been reported by respondents.

2.11

England, with 88% pre-1939 and 92% post-1945, dominates the UK in the two tables. Also, in the post-1945 set, the B1 units considerably outnumber any other type (see paragraph 1.26 above) and, as might be expected simply on cost grounds, the uncovered units (B, B1, B2, C) account for 69% of the total post-1945 - the pre-1939 proportion being approximately 50%. Type C streets - those shared by buses and pedestrians - form almost 10% of the post-1945 units. This may appear unremarkable but this proportion achieves significance when it is remembered that these streets are, more often than not, parts of network precincts. Furthermore, the 10% is a proportion of *all* units within which the malls of A2 units are included - a location clearly unsuited to large buses.

Clustering: the network precinct

2.12

The above discussion has little meaning for the shopper, for that person is less interested in 'units' - single streets, arcades or covered centres - than in clusters that form what we have termed the 'network precinct'. This type of precinct may be synonymous with the term 'place' unless the place has more than one precinct. Clustering of contiguous units seems to happen organically, perhaps with a little help from urban planners and property developers who see the advantages of locating within or adjacent to long-established shopping centres, whether pedestrianised or not. Ramsay (1977) suggests various desirable patterns of pedestrian routes, always assuming these can be achieved without radically restructuring a town centre. He favours a 'constellar' or 'polynuclear' pattern.

2.13

It is becoming quite commonplace for single streets to join up with others to form a network. Monheim (1975) shows that, for West Germany, outof 29 towns that had a single street precinct in about 1974, 83% expected to have a network precinct by the year 1980. In UK this occurs with many permutations - an example might be one conversion (B1) at the start of the process. Once this is seen to be successful in shopping turnover terms, a couple more might be converted, and linked

to the first (though this does not always happen). A nineteenth-
century arcade (A1) might add a further connexion, and it is quite
likely that a new enclosed centre (A2) could lead off the main street.
At this stage there might be complaints about poor access by bus, when
one of the earlier streets, or an additional street, could become a
type C - though many will also complain about *this* as an infringement
of their walking space.

2.14
Illustr. *II.11-II.12* show patterns of growth for Coventry and Swindon,
while the cover diagrams show the growth of the Chester network.
Swindon is an Expanding Town, so its centre has grown in parallel with
the growth in its population. Coventry was severely bomb-damaged
during the second world war; during its rebuilding it was in the
vanguard of thinking at that time by incorporating purpose-built
pedestrian precincts: the whole network has been consistently extended by converting existing streets to pedestrian precincts. Chester
probably has the earliest extant pedestrian shopping facilities (if
you discount the traditional footway alongside a busy carriageway),
which are doubtless among the earliest in Europe - its Rows. They date
from the thirteenth century in part and their upper level is covered
by the overhang of the remainder of the building. In 1981 the carriage-
ways of the Roman streets containing the Rows have had severe traffic
restrictions placed upon them.

2.15
Chapter 10 tabulates the information about precincts at the place or
network-precinct level, as well as at the unit level. It lists the
UK first tier local authorities, the number of Districts within each
of them known to have some form of precinct, the number of places,
within the Districts, that have a scheme, and the number of known
units within the County or equivalent area.

Other taxonomies

2.16
Two other forms of classification should be mentioned. One is concerned with space-sharing: a precinct could be classified by who uses it
(people walking; people walking and emergency traffic; people walking, emergency traffic and buses; the last three with access or with
service vehicles; and so forth). This aspect has been discussed by
Dalby (1976), who earlier (1973) reviewed the layout of shopping centres when related to pedestrian movement.

The second taxonomy concerns time-sharing. There are many cases where
precincts are reserved exclusively for pedestrians during peak shopping periods, and are turned over to service, access or all other vehicles outside these periods. This concept was briefly explored for
Putney High Street London, for peak late-night shopping and for Saturdays, in TEST (1976).

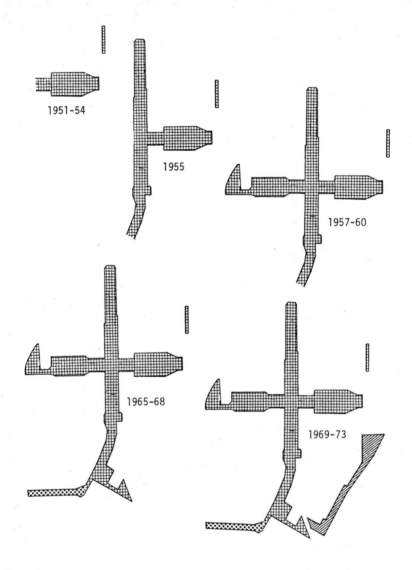

1951-54

1955

1957-60

1965-68

1969-73

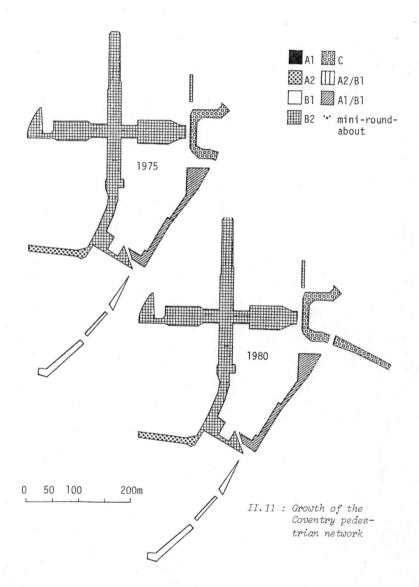

▓ A1	▦ C		
▩ A2	⫿ A2/B1		
☐ B1	▨ A1/B1		
▦ B2	∵ mini-round-about		

1975

1980

0 50 100   200m

II.11 : Growth of the
Coventry pedes-
trian network

39

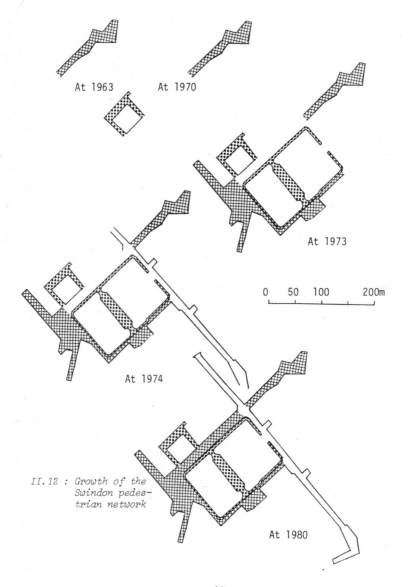

At 1963   At 1970

At 1973

0   50   100   200m

At 1974

II.12 : *Growth of the Swindon pedestrian network*

At 1980

2.17
This is a difficult question to answer, simply because suitable performance measures are not easy to apply across a cluster of nations. In
order to provide *some* information, each County of England and Wales is
listed in Illustr. *II.13* against the population / unit and the area in
hectares / unit, while the Counties themselves are arranged in descen-
ding order of density of persons per hectare. It is very important to
note that there are gaps in the information about the total number of
units. Where there are such gaps, the population:unit and area:unit
have been adjusted to allow for the estimated number of units in the
County, referred to in paragraph 2.06 above. When all these data were
tabulated, overall rankings for England and Wales for 'performance'
on the population:unit and area:unit bases, were derived.

2.18
Ranking all 54 Counties in that way gives a very biased result because
the catchment areas for precincts differ widely from the highly dense
conditions of the Metropolitan Counties to the sparsely populated deep
rural Counties. Partly to overcome this bias, the Counties were then
divided into broad groups with similar characteristics - Metropolitan
and Highly Urbanised; Medium Urbanised; Rural and Deep Rural. Rankings
by Group were then obtained and averaged to provide a final ranking
within each Group.

2.19
There is a great deal of information in Illustr. *II.13* What conclu-
sions may tentatively be drawn? In the first Group, West Midlands
seems to be a clear leader, Greater London ranks 6, and South Yorkshire seems to perform least well, despite the many streets that have
been converted in its County Town, Barnsley. In the Medium Urbanised
Group, Cheshire is again a clear leader; this is interesting because
its achievements are little publicised. Hertfordshire and Lancashire
share second ranking and Mid Glamorgan appears to perform least well
despite the considerable achievements of Cardiff. In the Rural Group
Durham is unchallenged except by the Isle of Wight, though they differ
widely in absolute population and area. Warwickshire, with several
historic towns, surprisingly performs least well in this Group. In
the Deep Rural Group North Yorkshire leads, because York has recently
acquired an elaborate network.

2.20
Chapter 2, supported by Chapter 10, has attempted a comprehensive
display of the types, timing and geographical distribution of pedestrian precincts in Britain. It also tries to inform the reader about
the qualitative and quantitative performances of County or equivalent
level authorities throughout Britain, though most of the discussion
centres around England and Wales, for the most data are held about
those countries. Using various measures, the more precinct-conscious
Counties are identified.

	RANKINGS BY ALL ENGLISH+WELSH COUNTIES					RANKINGS BY GROUP			
1	2	3	4	5	6	7	8	9	10
						Ran-	Ran-		Revised ranking within Group
County	Density pers/ha	Populn. per unit	Ranking on 3	Area per unit	Ranking on 5	king on 3	king on 5	7+8 / 2	
METROPOLITAN / HIGHLY URBANISED									
Greater London	46.48	69 951	49	1 505	2	9	2	5.6	6
West Midlands	30.99	27 624	9	891	1	1	1	1.0	1
Merseyside	25.57	59 250	44	2 316	5	8	5	6.6	7
Tyne & Wear	22.38	43 178	27	1 929	3	4	3	3.5	2
Greater Manchester	21.18	46 220	33	2 182	4	5	4	4.5	4
West Yorkshire	10.06	36 017	20	3 580	6	2	6	4.0	3
Cleveland	9.72	37 800	21	3 890	8	3	7	5.0	5
South Glamorgan	9.36	48 739	39	5 204	10	6	8	7.0	8
South Yorkshire	8.42	54 792	43	6 506	14	7	9	8.0	9
MEDIUM URBANISED									
Avon	6.74	45 100	32	6 693	15	10	4	7.0	6
Hertfordshire	6.45	32 928	14	5 101	9	3	2	2.5	2=
Surrey	5.92	70 071	50	11 827	22	18	11	14.5	14=
Mid Glamorgan	5.20	75 805	51	14 564	29	19	16	17.5	19
Nottinghamshire	5.00	48 650	38	10 828	20	13	9	11.0	11=
Berkshire	4.93	34 444	18	6 981	16	6	5	5.6	5
West Glamorgan	4.56	46 483	34	10 200	18	11	7	9.0	9
Lancashire	4.40	26 820 (24 382)	6	6 091 (5 537)	12	2	3	2.5	2=
Bedfordshire	3.75	69 555	47	18 651	33	17	17	17.0	16=
Kent	3.73	93 066 (38 778)	22	24 899 (10 375)	19	7	8	7.5	7=
Cheshire	3.72	13 730	2	3 689	7	1	1	1.0	1
Essex	3.68	43 677	30	11 860	23	8	12	10.0	10
Hampshire	3.63	38 059 (33 415)	15	10 486 (9 209)	17	4	6	5.0	4
East Sussex	3.61	50 000	40	13 820	26	14	15	14.5	14=
Staffordshire	3.54	321 000 (48 154)	37	90 610 (13 591)	24	12	13	12.5	13
Derbyshire	3.36	68 154	46	20 254	35	16	18	17.0	16=
Gwent	3.20	146 634 (43 990)	31	45 888 (13 766)	25	8	14	11.0	11=
Leicestershire	3.12	133 166 (66 583)	45	42 582 (21 291)	37	15	19	17.0	16=
West Sussex	3.02	33 889	17	11 208	21	5	10	7.5	7=
RURAL									
Isle of Wight	2.86	18 167	4	6 354	13	2	2	2.0	2
Buckinghamshire	2.53	79 333 (43 273)	29	31 404 (17 130)	30	14	5	9.5	9
Durham	2.49	13 217	1	5 300	11	1	1	1.0	1
Humberside	2.38	69 833	48	29 288	44	17	14	15.5	17
Warwickshire	2.30	76 000	52	33 035	47	18	17	17.5	18
Dorset	2.08	50 273	41	24 146	40	16	11	13.5	14
Northamptonshire	1.97	46 800	35	23 692	39	15	10	12.5	12
Oxfordshire	1.93	100 800 (42 004)	25	52 269 (14 333)	27	11	3	7.0	6=
Dyfed	1.83	157 226 (31 445)	13	85 995 (17 199)	31	8	6	7.0	6=
Gloucestershire	1.75	30 867 (27 235)	8	17 602 (15 531)	29	4	4	4.0	3
Cambridgeshire	1.48	36 071 (28 056)	10	24 369 (18 954)	34	5	8	6.5	5
Clwyd	1.47	35 816	19	24 264	41	9	12	10.5	11
Hereford and Worcester	1.43	43 231	28	30 229	45	13	15	14.0	15
Suffolk	1.41	24 409	7	17 288	32	3	7	5.0	4
Wiltshire	1.40	28 588	11	20 491	36	6	9	7.5	8
Devonshire	1.16	64 000 (40 727)	24	48 001 (30 546)	46	10	16	13.0	13
Norfolk	1.16	31 200	12	26 794	42	7	13	10.0	10
Somerset	1.12	43 000	26	39 542	49	12	18	15.0	11
DEEP RURAL									
Shropshire	0.96	33 700	16	34 929	48	3	3	3.0	3
Lincolnshire	0.85	35 928 (23 952)	5	42 070 (28 047)	43	2	2	2.0	2
Cornwall	0.81	53 909	42	66 450	50	6	4	5.0	4=
North Yorkshire	0.77	17 971	3	23 202	38	1	1	1.0	1
Cumbria	0.70	47 600	36	68 138	51	5	5	5.0	4=
Gwynedd	0.57	109 947	54	193 539	53	8	7	7.5	7=
Northumberland	0.56	40 000	23	71 956	52	4	6	5.0	4=
Powys	0.19	98 765	53	508 136	54	7	8	7.5	7=

II.13 : *An approximate guide to the performance of English and Welsh counties in the introduction of pedestrian units. Incomplete data are held about Counties with figures in brackets, which represent estimated population and area:unit ratios. Estimated data are used in the rankings.*

42

3
applying cosmetics

*The dimensions of unit- or network-precincts, and their environments,
are the concerns of this Chapter.*

*A mean width of street, from an admittedly small sample, of about
20 m, and a mean length of unit, from a much larger sample, of 150 m,
are deduced.*

*Various studies that have grappled with the problems of qualitative
assessment of environments are looked at. The many types of improvement associated with most precincts are then examined, and they include paving, seats, special lighting, vegetation and so forth. The
Chapter ends with a review of reductions in noise and air pollution
that have almost universally occurred after motor traffic has been
removed from a shopping street.*

III.01 (top left) Watford High Street III.02 (top right) Durham:
Silver Street and Castle III.03 (centre left) Newcastle upon Tyne:
Eldon Square with its shopping centre at rear III.04 (centre right)
Exeter Guildhall Centre III.05 (bottom left) Corby Shopping Centre
III.06 (bottom right) Durham, Milburngate Shopping Centre

3.01
The title of this Chapter is neither pejorative nor cynical. Any
urban area, or a shopping street in our case, can function perfectly
adequately for pedestrians if nothing other than the removal of vehicular traffic is carried out. Similarly, people are attracted to
other people for a variety of reasons, and may not be in the least
deterred by seeing a face untouched by applications. However, if one
adds trees, paving patterns or sympathetic lighting in the one case,
and male or female skin treatment in the other, the object of your
attention will usually have an enhanced appearance. Certainly, many
people, among them architects and town planners, see pedestrianisation as an opportunity for an essential face-lift (see Illustr. *III.01*).
Conservationists will also be involved, and those with responsibili-
ties toward historic environments (*III.02-04*). Shopkeepers often need
little encouragement to improve their facades, for this should stimu-
late better trade as well as adding generally to the environmental
quality,.

3.02
But enlivement of a town centre or other shopping area is not just a
matter of conserving the old. It has also been achieved through the
addition of quite new centres. These vary in architectural quality,
from hard-sell to hyper-sensitive, from Habitat intimacy to one-stop
shopping in the equivalent of a roofed-over football stadium. Some,
it is true, seem to have been constructed for short-term gain (*III.05*)
while others' materials will look well for a much longer time (*III.06*).
Again, environment is not just buildings and movement space - it also
includes the surrounding air, and later in this Chapter changes in air
pollution and noise will be discussed.

3.03
Several overviews of the urban design aspects of walking-and-shopping
places have been written. Whyte (1980) exposes individual responses
to small American urban spaces and provides many lessons on how to meld
street furniture, buildings and spaces into outdoor living rooms for
large families. The Design Council and Royal Town Planning Institute
(1979) had a rather different remit: they investigated the various
trappings customarily added to a street when it is pedestrianised.
These receive litter, are sat upon, provide a home for plants, illum-
inate the surroundings, help one to find things, accept parked bicycles,
enclose spaces, even advertise goods and local events. The Design
Council maintains a street furniture catalogue and its publication
'Street Scene' (The Council 1976) shows many well-designed pedestrian
precincts. Some forward thinking springs out of Garbrecht (1976) who
considers fundamental design requirements for people walking in both

town centres and suburbs. Edminster and Koffman (1979) discuss the environmental aspects of three US transit malls - streets shared by buses and pedestrians in British parlance - in Philadelphia, Seattle and Minneapolis; Pushkarev and Zupan (1975) also discuss the pedestrian's environment. Brambilla and Longo (1976a) illustrate the many adventurous features in US malls, ranging from a tent-like shelter, concrete sculptures and water in Sacramento, via lavish vegetation in Riverside Calif., to elaborate lighting in Evansville Indiana. They also show the plastic canopies over one of Wuppertal's streets in Germany.

3.04
Chapter 2 noted that precincts vary from single streets to networks. They also vary in scale - the width of component streets, the length of the precinct, and the height of flanking buildings. In the case of new covered centres all dimensions will be important, in two different ways - externally, where the physical bulk will matter very much in the urban environment and internally, where there are usually dimensions better related to the people using its malls. The table below gives mean wall to wall widths of 18 pedestrianised streets in 7 Districts (though note the small size of the sample):

District	No. of streets (units)	Mean width m
Barnsley	2	25
Birmingham	3	25
Bournemouth	2	12.5
Durham	8	10.5
Exeter	1	18
Newcastle-upon-Tyne	1	25
Wandsworth	1	22.5
All	18	19.8

III.07 : Widths of streets in 7 English Districts (18 units)

Whyte (see paragraph 3.03) tends to support small spaces, for he emphasizes (and his photographs support this) that people like to be with people: in large spaces there is not the same incentive for close encounters. Rarely in Britain do we have spaces of a size where a few people become lost, and Illustr. *III.07* seems to underline this: in a mean width of 19.8 metres this is unlikely, especially when this width has been effectively reduced by street furniture and planters. Interestingly, despite many people's views that North American streets are generally wider than British ones, the mean width of 69 USA pedestrian malls is 20.75m (derived from Brambilla et al 1977), but note again the small British sample.

3.05
There is great variation in the length both of units and of network

precincts. The table below explores the length of 825 known post-1945 units in UK by type of unit, as explained in paragraph 2.07 above: that covers all except the B1-C category, which simply means a unit or converted street part of which permits buses as well as pedestrians. The findings of Illustr. *III.08* are unexceptionable, though the small sample for Scotland has produced a distorted overall average length, whereas Northern Ireland's overall average is distorted by the 75% proportion of all its schemes being in the B category. From Brambilla et al (1977) the mean average length of 70 USA units is 443 m, almost three times the UK average. This is not surprising considering that the US store is generally much larger than its UK counterpart, and helps to explain why Denver has a US Urban Mass Transit Administration contribution of $33 million toward its pedestrian and transit mall. This is 7 blocks in length and will have small shuttle buses (TE&C 1979).

| Type of unit: | average length, m frequency of units of known length | | | | | Row total: Average units |
B1-C	A2	B	B1	B2	C	
England 314.0 5	226.2 114	147.8 61	118.9 353	133.0 113	196.0 93	151.0 739
Wales 200.0 1	120.5 4	95.0 2	124.9 22	100.7 3	396.7 3	146.0 35
Scotland	240.0 4	312.7 3	213.8 4	183.8 4	375.0 2	249.3 17
N Ireland	30.0 1	72.3 15			167.5 4	89.3 20
Channel Isles		60.0 2	113.6 11	190.0 1		111.4 14
Col. total: Average units 295.0 6	221.6 123	136.7 83	120.1 390	134.3 121	204.3 102	150.7 825

III.08 : Lengths of known unit precincts in UK, since 1945, by type of unit

3.06
Environmental quality is notoriously difficult to quantify. The Roskill Commission on the Third London Airport were one of the first teams to discover this, when attempting to put a monetary value on an historic church. Work by Nairn, Hellman, Goodey, Cullen (1971) and others in the UK has tried to remedy the situation, as have Appleyard and Lintell (1970, 1975) and Lynch (1960) in the United States. TEST (1976) while

studying pedestrian movement in Central Birmingham asked office workers
to record their rating of various environmental aspects of the pedestrian network, and of unchanged all-traffic streets, during a lunch-
time walk. Some of the results are debatable, or have too complex
qualifications to be quoted here.

3.07
However, Illustr. *III.09* shows the results from one exercise. Respondents were allocated 100 points to distribute between aspects of streets
they considered particularly pleasant for pedestrians. The more important aspects would receive the most points, provided the 100 were
not exceeded. In the table, 1.0 is least important.

Overall rank	Aspect	Overall mean score	Men	Women	Factor weight
1	Things to look at such as shops, posters, kiosks etc	16.1	16.3	15.9	2.5
2	Freedom from crowding on footways	14.5	13.3	15.1	2.2
3	Freedom from traffic	13.1	14.0	12.5	2.0
4	Clean, litter-free streets	11.9	10.5	12.7	1.8
5	Presence of seats, flo-wers, trees, shrubs	11.7	12.8	11.1	1.8
6	Level, even footways, free from obstruction	9.0	8.5	9.3	1.4
7	The opportunity to see and enjoy the quality of buildings and monuments	7.1	8.2	6.4	1.1
8	Ease of crossing street	6.5	7.2	6.0	1.0

*III.09 : Environmental scores and weightings for Central Birmingham
streets*

The results, from a sample of about 200 people, cannot be regarded as
definitive for Birmingham, let alone the rest of Britain, for persons in
different occupational groups, as an example, are likely to have different priorities. Nevertheless, in this particular case there seems
to be a strong reaction against uninteresting environments, and strong
support for the types of environmental improvement that are frequent in
pedestrian precincts.

3.08
When individual cosmetic treatment is investigated, a great deal of
evidence is available. For reasons mentioned above, no attempt will
be made to quantify this, nor to analyse it comparatively for the large
number of UK precincts. A select sample of 27 places is provided in

Illustr. *III.10.* Nevertheless, for the majority of entries in the Chapter 10 tabulation there will be some documentation on environmental aspects of each precinct - interested readers could contact the District Council concerned.

Place	Aspect considered	Source
Barnsley	Retrospective and predictive views of Central Area streets	Barnsley MBC (1977)
Birmingham	Comprehensive	City of B'ham (n.d.)
Bolton	Floorscape, photographs, responses	Stewart et al (1979)
Bristol	do do do	do
Cardiff	Floorscape mainly	Surveyor (1977)
Chichester	Comprehensive	West Sussex CC (1976)
Darlington	Responses to Survey	Darlington BC (1980a)
Dundee	do do	City of Dundee (n.d.)
Durham	Floorscape	Scott (1979)
Glasgow	Comprehensive	GLC (1972)
Gloucester	Photograph	Perkin (n.d.)
Hereford	Floorscape, photos, responses	Stewart et al (1979)
Kings Lynn	Photograph	Perkin (n.d.)
Leeds	Comprehensive	GLC (1972)
Leicester	Lack of improvement	London Transport (1974)
Newcastle/Tyne	Furniture, layout	do
Newham	Comprehensive	Lund (1974)
Norwich	Photograph	Perkin (n.d.)
Oxford	Floorscape	London Transport (1974)
Portsmouth	Responses to survey	Ove Arup (1973)
Reading	Pavements widened, only	London Transport (1974)
Salisbury	Photograph	Perkin (n.d.)
Stevenage	do	do
Wandsworth	Comprehensive	Wandsworth LB (1979)
Westminster		
Carnaby St	Comprehensive	Myatt (1975)
Leic. Sq.	do.	GLC (1978)
Oxford St	do.	Parker & Eburah (1973)

III.10 : Some environmental documentation for precincts in 27 places

Noise

3.09
Some before and after pedestrianisation noise levels for a variety of locations are provided in Illustr. *III.11.* The table shows some dramatic, and some moderate, reductions. Even the relatively small difference at Uppsala (of 4.1 dB(A)) is perceivable by most people, however. A great difference would be perceived in Cologne, with its 25 dB(A) reduction.

While not providing the same degree of detail, OECD (1978) shows that in 105 OECD-area cities, noise levels were reduced in 86% of them.

Location	Noise level, dB(A) Before	After	Source
Cologne	70	45	1
Copenhagen	75	65	1
Gothenburg	74	67	1
Leeds	75	65	1
New York	75	65	1
Exeter High Street	82.6	72.5	2
Uppsala	77.5	73.4	2
Sutton High Street	80	74	3
Chichester, North Street	72.5	61.5	4

III.11 : Changes in noise level before and after pedestrianisation
Sources: 1 Brambilla and Longo (1976a); 2 TEST (1981);
3 TEST (1980); 4 West Sussex CC (1976)

Air Pollution

3.10
Air pollution caused by vehicular traffic is a complex subject because of the number of pollutants involved, because of their interactions with themselves and with other chemicals, and because their location and concentrations are related to micro-climate, which varies through-out a 24-hour day and by different times of the year. It is therefore rather less easy to provide good evidence of reductions in certain pollutants as a result of reducing or removing traffic, than it is to discuss reductions in noise levels. For some of the problems in-volved see Open University (1975) and TEST (1975). Some changes are tabulated in Illustr. *III.12* on page 51.

Location	Pollutant, unit	Before, or control location	After	Source
Philadelphia	CO ppm	6.98-8.28	2.98	1
Gothenburg	CO ppm	65	5	2
Vienna	'air pollution'		40% less	2
Cologne	lead mg/m^3	4	1	2
New York	CO ppm	23	8	2
Tokyo	'air polln.'ppm	8	2.9	3
Marseilles	do	14.3	3.4	3
Durham	lead $\mu g/m^3$	0.83	<0.1	4
	smoke $mg/100m^3$	22.6	<4	

III.12 : Changes in air pollution levels before and after pedestrianisation
Sources: 1 Edminster and Koffman (1979); 2 GLC (1972); 3 Brambilla and Longo (1976a); 4 TEST (1981)

Illustr. *III.12* shows substantial reductions in pollutants in all the cities mentioned, though the cautions earlier in this paragraph should be noted.

The OECD document referred to in paragraph 3.09 also states that there was 'a marked decrease in air pollution in 72% of the 105 cities.' For more technical information on air pollution and pedestrianisation see Sainsbury and Caswell (1977).

3.11
This Chapter has considered the dimensions of unit- and network-precincts, dwelt upon the environmental improvements that normally go along with the creation of a precinct - paving, lighting, seats, litter bins, vegetation, special signposting, and so forth- and finally it has looked at the reductions in noise and air pollution that are found, not surprisingly, to accompany the substantial removal of motor traffic. On page 52 some more photographs are provided to illustrate the theme of environmental improvement.

III.13   III.16
III.14   III.17
          III.18
III.15

III.13: Sutton High Street
III.14: Sunderland shopping
precinct III.15: Regent St.
Swindon III.16: Princesshay
Exeter III.17: Commercial Rd
Bournemouth   III.18: Lymington
Hampshire

52

4

getting there

Chapter 1 introduced us to this topic. This Chapter discusses it in more detail, and includes those means of transport omitted earlier - for moving both people and goods. Whether a pedestrian scheme went ahead or not has often depended on finding satisfactory ways of delivering goods to shops.

Another determinant has been the accommodation of traffic displaced by a pedestrian scheme, though West Germany and North America are less worried than the UK. Still, the UK does have a few iconoclasts who believe nothing needs to be done, and that this traffic tends to disappear (the obverse of it appearing in great numbers when a new road is opened).

For many retailers and suppliers, servicing anywhere other than at the shop frontage within time or weight restrictions has been regarded as crucial; of the available methods, rear, above or below servicing are preferred. However, in West Germany there is nothing like the same concern: a far higher proportion of their shopping precincts are serviced from the pedestrian streets, usually outside peak shopping hours.

IV.01 : Several ways of reaching Sutton High Street

4.01

The accessibility of shopping precincts to individuals depends on their knowledge of what is there and how to reach it; on the availability of time and money; on the availability of some form of transport; and on any barriers such as psychological distaste for public places or fear of assault. Jones (1981) in fact defines accessibility as being 'concerned with the opportunity available to an individual or type of person at a given location to take part in a particular activity or set of activities.' Jones' literature review is a good starting point for those wishing to know more about accessibility. Among some of the obvious, and some of the interesting, phenomena he quotes are the following:

* OECD (1977) suggested that, as accessibility for shopping and personal activities increases, the overall trip rate for these activities rises, the number of public transport trips falls, and the number of car trips first increases and then declines to about the original level
* Vickerman (1974) found tentative evidence (for the not particularly surprising conclusion!) that accessibility to shopping and leisure facilities affected the numbers of trips made for these purposes
* Thiebault et al (1973) suggested that access to activities such as primary school or shopping centre may be as important as access to the workplace

    and that

* Richer people tended to have better accessibility than poorer people....

4.02

Another TRRL publication on accessibility is Mitchell and Town (1977) and, for an example of work in County level transportation studies, that for West Yorkshire has been reported by Cooper et al (1979); this article usefully tables the accessibility standards adopted. Two publications by TEST (1980, 1981) are particularly relevant: the first records a study of the trade-off between accessibility and environment in the London Borough of Sutton, while the second reviews accessibility by bus to pedestrian precincts.

4.03

A wide range of people, vehicles and goods have to reach and leave pedestrian precincts. This Chapter considers

- self-powered people (walkers, cyclists)
- individual or small group vehicles (mopeds, motorcycles, cars, light vans, taxis)
- buses and rail-borne transport

essentially as passenger transport; goods, maintenance and furniture removal vehicles, and construction traffic, come under the heading

- servicing traffic

while other types of transport serving particular roles are

- disabled persons' vehicles
- emergency vehicles (police, fire, ambulance).

4.04
All of these means of transport have to be accommodated in the design of a pedestrian precinct scheme: some of them will be permitted within it while others will be displaced onto other parts of the town's transport networks, whether existing or specially built. In order to comprehend the wide variations in means of transport used to reach pedestrian areas, a number of 'modal splits' are tabled as Illustr. *IV.03*; the variations are not arbitrary, of course. They result from different levels of car ownership, quality of public transport, number of people living within walking distance of the shops, car parking facilities, and many other variables. First, let us consider North America where it might be assumed the car would dominate access to virtually all precincts. In Brambilla et al (1977) 66 USA cities with precincts are considered on car access and the results are as follows (the percentage not by car represents the total of bus, taxi and train users, those walking, and those cycling.)

% by car	No. of cities	Cumulative %
91 or more	12	100.0
81-90	29	81.8
71-80	13	37.9
61-70	4	18.2
51-60	2	12.1
41-50	3	9.1
40 or less	3	4.5
	66	

IV.02 : Percentage of those arriving by car at 66 United States precincts

Apart from the fact that Illustr. *IV.02* belies the commonly held belief that car use in USA is near universal, it is also interesting to note that 2 of the 3 cities with 40% or less car use are also ones with the highest populations in the 66 cities quoted: Baltimore (906 000 population, 33.3% car) and Philadelphia (nearly 2 million people, 36.9% car).

4.05

If we now turn to 'modal splits' for UK shopping journeys, some examples are given below:

	National sample 1975-6 NTS	Sutton 1979	Barnsley 1975 Peds, Sat'day	Birmingham 79	Durham 1979	St Johns Rd 1977 Sat.	Palm.Rd.P'mouth Before scheme	do.after scheme	Gateshead Sat. before scheme	do. after scheme	Kings Lynn 1975	Dundee 1975	Hammersmith 79	Wakefield 1976	Seaham Durham 73
Public transport	16	21.7	26	65	33.1	34	20	21	45.4	47.1	22	52	29	38	46
Car, van, lorry	35	43.4	58	19	41.5	19	38	37	27.4	25.0	60	32	17	38	27
Walk	46	32.3	16	15	21.3	47	37	37	27.0	27.3	15	14	50	26	27
Other private	3	2.6	-	1	3.9	-	5	5	0.2	0.6	3	2	4	-	-

IV.03 : Some modal splits for shopping journeys in Britain (see also
Illustr. VI.01)
Sources: Department of Transport 1979, TEST 1980, Barnsley MBC
1975, West Midlands PTE 1979, Wandsworth LB 1977,
City of Portsmouth 1976, Gateshead BC n.d., City of
Dundee 1976, Durham CC 1973, Davis Ives Assoc. 1977,
LB Hammersmith 1980, Norfolk CC 1975

Two aspects of the above table are notable. First, the wide variations about the national average quoted in the first column. Second, the slight changes in distribution of means of transport between the 'before' and 'after' conversions of two streets to bus and pedestrian only use. In Gateshead, the buses were reintroduced, after a time when the street was purely pedestrian, at the request of the street's users (Murray and Ennor n.d.): conversely, in the Leeds network, a minibus service was removed at the request of users of the pedestrianised network.

Walking and cycling

4.06

Illustr. *IV.03* shows the importance of walking, though does not distinguish cycling from 'other private'. While cycling is growing in popularity it still accounts for a relatively small proportion of the means of transport used to reach a precinct. Paragraphs 1.06-1.11 discuss walking to the shops in some detail.

Powered cycles, cars, light vans and taxis

4.07

IV.03 also does not distinguish powered cycles, though again the proportion is quite small. Light vans are also relatively unimportant and the use of taxis for shopping trips will vary between the type and economic status of the urban area. Private cars, on the other hand, constitute the highest proportion of the modal split in four of the 15 examples in *IV.03* - which is rather unexpectedly low. Naturally there is a large literature on parking, of which OECD (1980) is particularly interesting. This work attempts to evaluate urban parking systems in 14 world cities, studied by delegates from 12 countries, notably excluding Britain. National Car Parks provide lists of their car parks, which form a large proportion of those close to precincts. Illustr. *VII.02* provides data on types of car parking used by employees of businesses in a precinct, while Darlington BC (1980b) provides a thorough review of parking of all types that may be used by shoppers in its precincts.

Buses and railborne transport

4.08
These topics are well represented in the literature. Mention will only be made here of some of the more significant references. For a broad view the National Bus Company (1975a, 1978) have prepared two reports on Bus Priority Schemes, which include those streets shared by buses and pedestrians (though the reader will find that the Type C streets in Chapter 10 of this book represent a larger collection, mainly because of the difference in time), and NATO (1976) have published a report with a similar title to NBC's. Kerridge (1979) has evaluated priority schemes that have increased patronage. TEST (1981, 1980, 1979) reviewed buses and pedestrian areas, and looked particularly at the effects of pedestrianisation on buses in Sutton and Kingston, respectively. London Transport (1974) examined buses in pedestrian areas, Goddard et al (1977) produced an internal paper for Greater Glasgow PTE, with a clearly stated logic for buses in central areas, and the Greater London Council (1979a) produced a committee document on buses in town centres. The Department of the Environment undertook about ten Bus Demonstration Projects in the early 1970s, some of which involved precincts. Ball and Brooks (1976) looked at design requirements for buses in pedestrian streets, a topic which also in part concerns Bradburn and Hurdle (1981). Finally, we might note particular studies in three places: in Oxford (City of Oxford Motor Services Ltd 1978), Portsmouth (City of Portsmouth 1976) and Canterbury (National Bus Company 1975b).

4.09
Some data, on the location of bus stops related to Places (see paragraph 2.02 above), are available. These have been culled from County, District Council, and bus operator replies to requests for information. Illustr. *IV.04* distils this information; it relates to 59 places in 43 Districts or equivalent, themselves being in 21 Counties or equivalent. Average number of bus stops within each distance band are presented for the 59 places.

| | Number of bus stops (direct distance, m) | | | |
	0-50	51-100	101-150	151-200
Total	313	168	128	131
Average stops per place, n=59	5.3	2.8	2.2	2.2

IV.04 : The location of bus stops related to 59 places in 43 Districts

Taking the midpoint of each range, the average distance of a bus stop from a precinct, using the above data, is 80.2 metres. However, there are wide variations between network pi ...
bus stops horizontally or vertically adjacent (Birmingham and Newcastle upon Tyne are prime examples of this) at one extreme, and small market towns with perhaps five buses a day, at the other. Illustr. *IV.04* interestingly shows a preponderance of stops within the 0-50 m range.

4.10
In cities that have metro or more conventional rail systems these tend to be well used for shopping trips. In countries like Germany that also retained tramways when all around them were tearing up the tracks a wide range of public transport access to shopping precincts is available. In Britain London Transport has extended its underground network in recent years: the Victoria and Jubilee lines connect with important shopping areas at several stations. Tyne and Wear's metro in its first year of operation doubtless carried many shoppers to Newcastle's central pedestrian network, for three of its stations link with this. In Merseyside, Liverpool's precincts have been served by a rail loop and in Glasgow, the renewal of its underground railway has done the same for its precincts.

Servicing traffic

4.11
Providing good access for deliveries to and collections from shops seems to have been one of the major problems delaying the introduction of shopping precincts in Britain. Other countries do not seem to share Britain's obsession with totally segregated servicing - either from the rear of the shop, or to servicing bays under or over it, rather than to and from its frontage at offpeak shopping times, or completely outside those hours. Jennings et al (1972) studied the servicing of the Watford precinct, which is a mix of rear and frontage servicing. The trade federation, which has changed its name over the years, has published two substantial papers describing its policy and on the need for satisfying its particular conception of servicing

facilities (Multiple Shops Federation (1963) and British Multiple
Retailers Association (1980)). The latter supports the concept of pedestrianisation, though moderates this with 'three essential qualifications: a) prior consultation with affected traders to achieve, with
mutual consent, reasonable access for servicing; b) alternative routes
for public transport very close to shops, and c) adequate car parking
within 200m of the principal shops.' Unexceptionable, one might feel,
though some trading off with the increased turnover of the Association's
shops after pedestrianisation might be fair.

4.12
Perhaps the Germans have accepted that trade-off. Wood (1966) discusses,
for example, Schadowplatz in Dusseldorf none of whose buildings have
rear entry facilities; they are frontage-serviced outside the period
10.30-19.00. Monheim (1975) takes this much further. He had responses
from 125 cities and towns on where 'pedestrianised' shops were serviced:
76% were 'on the foot street', 16.8% were on side streets, 2.4% were from
yards or underground, and 4.8% were serviced by more than one method.
This seems far removed from the demands for rear, over or under, servicing. Clearly to service the frontage limits the size of servicing vehicle - the paving slabs used in precincts fracture easily under heavy
point loads - so either medium to small vehicles have to be used, or the
goods transferred from large lorry to a trolley at the end of, or within
a side street leading off, the pedestrian street. Presumably the Germans
have solved this too.

4.13
What types of servicing are to be found in the larger British sample investigated for this book? These have been broken down by time period and
precinct category. Pre-1939, for 119 units, 6 are serviced at the frontage at all times, 1 for restricted times at the frontage, 80 by trolley
(reflecting the fact that a large proportion of pre-1939 units are arcades,
too narrow to take vehicles), 6 from the rear, and 26 with various combinations of location. For the post-1945 period, information is held for
800 units, broken down as in Illustr. *IV.05*.

4.14
What is immediately apparent from Illustr. *IV.05* is that a total of
46.3% of the types occur within categories 4 and 5 - or rear, side,
above or below servicing. 115 of the 119 combination-of-servicing type
included part of types 4 or 5, and 24 were *solely* combinations of types
4 and 5. If these 24 are added to types 4 and 5, the total is 394 or
nearly 50% of the total sample (compare this proportion with the 19.2%
of West German schemes with 'side, yard, or underground' servicing:
paragraph 4.12.) Again, compare the 76% of German schemes served from
the foot street to the 34.63% of the UK units served in this way.

4.15
A rapid assessment might suggest that the British Multiple Retailers
Association's pressure for off (ped.) street servicing is proving very

	Frontage, unspecified	Frontage, unrestricted	Frontage, restricted	Trolley	Rear	Side, above, below	Combined	Row total
	0	1	2	3	4	5		
A		2			3			5
A1		1			2			3
A2				2	49	58	12	121
B	12	1	21	3	7	1	1	46
B1	17	73	90	20	82	21	61	364
B2	11	2	1	4	90	20	11	139
C		17	13	1	8	2	28	69
Combined	1		3	2	14	10	5	35
Dont know	1	3	8	2	2	1	1	18
Col. total	42	99	136	34	257	113	119	800
Col. %	5.25	12.38	17.0	4.25	32.13	14.13	14.88	100.0

IV.05 : Types of servicing, by precinct type, for 800 post-1945 UK units

effective. Clearly it is, although closer examination shows that it is the A2 and B2 schemes that dominate the off-street servicing categories. This is, perhaps, understandable, for these types are newly built schemes where it would be perverse not to service them in this way. On the other hand, the B1 schemes - streets converted from historic carriers of all kinds of traffic, and difficult to service other than from the front - show 28% with category 4 or 5 servicing (as against the nearly 50% for the entire sample), and nearly 50% serviced from the pedestrianised street.

4.16
Finally, some servicing arrangements and issues in particular places might be mentioned. In Chesterfield there were objections to various precinct proposals mainly because of the projected servicing arrangements: presumably these have been overcome, for the town centre in 1981, during various conservation works, seems to have the potential for a distinguished shopping centre. In Barnsley, 'servicing of shops is something of a problem, as there are few rear servicing facilities - on the two bus and pedestrian streets (as they used to be) other vehicles were permitted only between 18.00 and 10.00. Birmingham

places no restriction on frontage servicing where no rear servicing is
possible. In Bournemouth's two bus and pedestrian streets, Commercial
Road is serviced mainly from the rear, Old Christchurch Road from the
street. The arcades are serviced by trolley. Durham has a variety of
solutions for its network: some streets have 23.00-10.00 servicing;
some shops have rear service yards; others receive trolleyed deliveries, the lorry remaining in the Market Place; the City Council's own
enclosed precinct is serviced from alongside. Exeter's High Street has
rear servicing, while Newcastle's Northumberland Street is supposed to
be frontage-serviced from 17.30-11.30 though violations occur. (All the
above information except for Chesterfield from TEST 1981). Swindon's Brunel
Centre unusually has overhead servicing - see Illustr. *V.02*.

Disabled persons' vehicles and emergency vehicles

4.17
Most precincts make some allowance for disabled people and their vehicles
and are doubtless happy to do so for genuine cases. However, there are
many cases where the 'Orange Badge' is violated, a point discussed in
more detail in Chapter 5, and one major city is seeking ways of excluding
Orange Badge vehicles altogether.

4.18
All precincts have some arrangement for easy access by police, fire and
ambulance vehicles, though special methods have to be adopted for pre-
1939 arcades and for the recent enclosed shopping centre. There is one
final type of access that some authorities specifically regulate - if
there are dwellings within a shopping precinct then both removal vans
and hearses occasionally need access.

Displaced traffic

4.19
One issue that carries far more importance in Britain than in either West
Germany or the United States concerns what is done about traffic displaced
from a pedestrianised street. In Britain this usually means a new road
near to the pedestrianised area. While there are often other reasons
that also support this new road, it is interesting that:

- North American malls rarely have any new roads associated with them
  them;
- Only about 59% of the 146 German schemes in Monheim's book have
  a new road constructed nearby, which may not have been caused
  by the precinct;
- Wood (1977) believes that displaced traffic frequently disappears
  or does not add to the flows in surrounding streets post-precinct;
- Much central area traffic does not need to be there, as essential
  user studies by the Greater London Council and the London Borough
  of Hammersmith have shown.

5

being there

One of the humane reasons for creating pedestrian precincts is to reduce collisions between people and vehicles. There is consequently a large literature on accident prevention. There is widespread evidence, not unexpectedly, that accident levels reduce considerably after all, or most, motor vehicles are removed from shopping streets, though it is heartening to know that those streets retaining buses not only ensure a high level of accessibility by this means of transport, they also show high safety achievements.

Once a precinct is opened there will inevitably be those who abuse it, roughly of three types: those who deliver or collect goods outside the permitted period; prohibited vehicles which enter a precinct; abuses of the disabled drivers' 'Orange Badge' system.

The Chapter ends with some discussion of pedestrians' space requirements and changes in pedestrian flows after a precinct has been created: almost invariably there are large increases.

V.01 : Cattle pens, Waterloo Road London   V.02 : Canal Walk, Swindon,
a largely accident-free environment (this photo and III.15 courtesy
Thamesdown Borough Council)   V.03 : Walking and buses, no conflict,
High Street Watford   V.04 :   Need for pedestrian signals?   High Street
Watford.

5.01
This chapter is concerned with behaviour. While changes in accident rates resulting from pedestrian/vehicle conflict dominate the literature, there are some data on pedestrian flows, densities and street crossing. Ganguli (1974) has investigated pedestrian delay at locations in London. A moderate amount of material is also available on observance of precincts' regulations by non-pedestrians.

Accidents

5.02
In this International Year of Disabled Persons it is chastening to note that an estimated 30 million people in the world were disabled as a result of traffic accidents when the figures were published in 1976 (Anderson 1981). As these accidents concern most central governments, so references abound in most countries' literature. Thus Japan has a widespread policy of banning vehicle traffic for only a certain period of each day: a decrease in deaths and injuries has occurred (Murata 1978). TEST's bibliography (Elkington et al 1976) lists 68 *general* references on accidents;there are many more than that when specific factors are taken into account. OECD's Special Research Group on Pedestrian Safety (1977) has published a 3-volume report, and there have been reports on drinking and driving, driving while disqualified and even a 1973 Conference in Amsterdam on the Biokinetics of Impacts.

5.03
When we start to investigate changes in accidents involving pedestrians resulting from the creation of precincts, the number of references diminishes, though there will be many more reports and papers prepared by local authorities than the ones noted here. From a personal communication we can note that the bus and pedestrian street of Gallowtree Gate, in Leicester, experienced a reduction from 6 accidents a year to less than one a year, since the scheme was introduced. GLC (1972) quotes a 20% decrease in accidents since the Gothenburg scheme was implemented, while Edminster and Koffman (1979) discuss the moderate reduction in Chestnut Street in Philadelphia (from 25 to 24 pedestrian injury accidents per year); however, this is a transit mall, so buses do remain with pedestrians. In a roughly similar situation in Oxford, Dalby (1976) shows a rather larger reduction in accidents involving pedestrians in two bus and pedestrian streets: Queen Street and Cornmarket Street.

5.04
Some information on accidents is contained in TEST (1981). Various
case studies were undertaken and those commenting on accidents are as
follows. In the bus and pedestrian streets in Bournemouth accidents
are said to have decreased considerably (County Council) or not to
exist (Bournemouth Transport) since scheme introduction. 'The high
proportion of elderly retired people and the fact that "pedestrians
tend to take over road space requiring bus drivers to proceed with
considerable caution" make this a significant achievement.' In Durham
City all accidents involving personal injury reduced over the three-
year periods 1971-3 from 3 serious and 41 slight to none and 8 respectively in 1976-8. Personal injury from buses declined from 11 to 0.
In Exeter, all personal injury accidents decreased from a range of
8-14 a year 1969-74 to 4 in 1978, the only post-pedestrianisation year
with complete 12 months' data. In Uppsala traffic accidents reduced
by 47% in the reorganised streets and increased by 4-12% on adjacent
main streets. Finally, in LB Wandsworth the total personal injury
accidents in a year varied by 11-14, 1974-76 and reduced to 6 in 1978
and 4 in 1979; these results are particularly satisfactory for the
data relate not only to the street from which most vehicular traffic
had been removed, but also to its junctions with other heavily trafficked roads at each end. All the case studies above included a bus
and pedestrian only street within a network studied; in some cases
(Exeter and Wandsworth) the data relate only to the bus and pedestrian
street.

Violation of new regulations

5.05
Violation of precinct regulations occurs under a number of headings,
among which are:

- deliveries and collections outside permitted hours
- entry of vehicles that have been excluded from a precinct
- abuse of the Orange Badge system.

One cause of servicing outside permitted hours is that lorry or van
drivers do not want to work unsociable hours. Another is that deliveries by a large HGV are often made sequentially: its capacity is so
great that several shops in different towns can be serviced by the one
vehicle; not unnaturally some arrivals at shops will be at the wrong
time. Whatever the cause, a visit to any of a large number of precincts
will show unauthorised deliveries or collections taking place, which
will continue as long as there is inadequate enforcement (or in the
view of the British Multiple Retailers Association (1980), there are
inadequate facilities for servicing). Rather similar comments can be
made about the illegal presence of certain other vehicles within
precincts, though in this case there appear to be fewer violaters.
Nevertheless London's Oxford Street has a large number of illegal
private cars at any particular time and this arouses the wrath not
simply of pedestrians, but apparently of the shopkeepers too.

66

5.06

A number of local authorities assert that there is abuse of the 'Orange Badge' system for disabled persons' vehicles: either by the badge-holder or by another person using the vehicle who is not disabled. Local authorities in both Durham City and Newcastle-upon-Tyne expressed concern at the number of Orange Badge parkers within precincts. However, better evidence is contained in a personal communication (Robertson 1980) which provided details of a survey in the City of Derby, on a Thursday and Saturday in March 1979, in East Street and Tenant Street. The broad findings were that:

- many vehicles were parked totally illegally;
- only a small percentage of vehicles were parked completely legally;
- many people using the Orange Badges have no apparent physical handicap, which impairs mobility, whereas the Disabled Persons' (Badges for Motor Vehicles) Regulations 1975 clearly state that the scheme is intended for persons having considerable difficulty in walking.

Pedestrian space requirements

5.07

Pushkarev and Zupan (1975) provide much information on theoretical space requirements and cite the work of other authorities on this topic. They note the problems of deciding what space a 'standee' requires: this varies from 0.09 m^2 for a female sardine, packed vertically on a footpath, to 0.94 m^2 for a man with an opened umbrella. But these do not allow for movement, nor for the avoidance of hazards ('Thus, bumping into an object, or making a violent maneuver to avoid it, is an indication of crowding.') Another important factor is 'the intensity with which (people) are communicating, and differs from culture to culture.' Fruin (1970) suggests levels of service for standing pedestrians in Illustr. *V.05*:

Quality	Spacing, m	Area/person, m^2
Unimpeded	1.2	1.2
Impeded	1.0-1.2	0.7-1.2
Constrained	0.6-0.9	0.3-0.7
Congested	0.6	0.2-0.3
Jammed	0	0.2

V.05 : Levels of service for standing pedestrians

5.08

'Flow' is defined as

$$flow = speed \times density$$

67

but Pushkarev and Zupan prefer a measure which is the reciprocal of density, available space per pedestrian, whose equation is:

$$\text{space } \frac{m^2}{ped} = \frac{speed \ (m \ / \ min)}{flow \ (ped/min/m)}$$

The authors go on to quote Oeding (1963) who very sensibly says that shoppers are the most inefficient walkers, attaining perhaps only two thirds of the flow at three quarters of the speed (workers leaving manufacturing plants are the most efficient!) To complete this context for understanding data on pedestrian flow, Older (1968) suggested that the maximum flow of pedestrians per metre of walkway was just over 76 persons per minute and each pedestrian would require 0.51 m^2. One final point on the efficiency of pedestrian movement as a utilizer of space: Pushkarev and Zupan note that between 08.00 and 20.00 'autos and taxis produce, on almost the same amount of space, less than half the person-miles of travel that pedestrians do.'

5.09
One reason, not mentioned previously, for creating precincts is that pedestrian flows and densities on footways adjacent to shops are so high that they constitute 'crowding', the result being that people are forced on to the carriageway to mingle with vehicular traffic. This problem influenced LB Hammersmith in their wish to pedestrianise King Street, though this has not yet been accomplished. Garton (1977) while studying the effects of pedestrianisation in Barnsley, obtained before and after responses from several hundred pedestrians on crowding. A 21-point attitude scale was used where 1=Bad and 21=Good. 'Before' in Market Street had a mean response of 8.78 while 'After' in Market Street and Queen Street had 11.51.

5.10
The effect of pedestrianisation on pedestrian flows can be illustrated from GLC (1972) where it is said that Dusseldorf's and Munich's flows were 60% greater after the event, and that in the New York experiment (Madison Avenue) there was an increase from 9000 to 19 000 pedestrians/hour. Edminster and Koffman (1979) note that Minneapolis actually experienced a distinct decline, which was partly explained by the development of a skyway system by which many stores are connected by bridges across the Mall at second floor level. In Philadelphia there was initially a significant increase, which then stablised at a point above that pre-transitway. In Durham (TEST 1981) pedestrian flows have increased on average by 29%, though Elvet Bridge shows an increase of about 47%. Parker and Hoile (1975) note that London's Carnaby Street experienced a 36% increase in the number of pedestrians between the period before the scheme was introduced and after the traffic was removed - but before the street was repaved.

5.11
A report on its larger town centres' pedestrianised areas was produced by Greater Manchester MCC (1976); this contains information on

pedestrian flows versus pavement flow/capacity on a weekday and on
Saturday, on crossing flows and average delay on the crossings.
Needless to say this cannot be reproduced, or even summarised, here,
but does show the admirable detail that some local authorities consider when planning their town centres. Another document with some
detail of pedestrian flows forms part of the Darlington Central Area
Plan reports (Darlington BC 1980). In Norwich a pedestrian flow
survey was carried out in 1968, one year after opening the converted
London Street: in a nine-hour period there was an increase of 11 339
people over the 'before' condition (Wood 1969). Finally, Newcastle-
upon-Tyne has experienced considerable shifts in pedestrian flows
since the opening of the Eldon Square Centre, and these shifts are
discussed in Bennison and Davies (1977).

5.12
In Seaham, County Durham, data on number of pedestrians entering
Church Street before and after closure to traffic are available
(Durham CC 1973). 1595 entered the street before closure. Immed-
iately after closure this was 1088 - an unexplained drop - which then
improved over the next six months to reach 2074.

5.13
This Chapter has discussed the way people behave in pedestrian precincts. Sometimes this has meant fewer accidents involving pedestrians (partly because damaging vehicles have been removed, partly
because those that are allowed to remain have taken extra care when
mingling with walkers). The flow of people walking has normally
increased, usually by quite large numbers - though it is worth remem-
bering that precincts do not automatically mean large increases in
space for walking because part of the newly gained space quickly
fills up with seats, trees, flower beds, signs, lights and so on.
The other change in behaviour manifests itself as abuse of the regulations, and Illustr. *V.06-V.08* on page 70 show a few examples of
what can happen.

69

6

responding to change

People's attitudes toward pedestrian shopping schemes have frequently been monitored through various types of interview survey, and such methods have less often been applied to shopkeepers, transport operators and their crews, and other interest groups.

The descriptions of these surveys fall into one of several types. There are international ones, sponsored by agencies like OECD, who ask many city managements their views about precincts. There are County-level surveys that sometimes cover more than one location within their authority. And there are individual, local surveys, often undertaken either by a District Council or by a University or consultancy research team.

The results of many surveys are sampled in this Chapter. One overwhelming response is the high level of approval for the creation of shopping precincts, mainly from their users, but often from all groups who are interested in the scheme.

LONDON BOROUGH OF WANDSWOR

CONFIDENTIALITY OF RESULTS

THE INFORMATION YOU GIVE IN COMPLETING THIS FORM WILL BE TR⌐
CONFIDENTIAL AND USED ONLY TO PRODUCE TOTALS FOR AREAS ⌐⌐

(vertical text, left margin) ⌐ENT STUDIES: IMPROVING THE PEDESTRIANS' ENVIRONMENT

(vertical text) loyee spending most of your time working outside this building pleas⌐
n in the collection box provided (DO NOT COMPLETE THE QUEST⌐

(vertical text) udies have been asked by the Department of the Environment to und⌐
y the needs of pedestrians so that in future better facilities might b⌐
us by completing this short questionnaire about walking in this are⌐
STIONNAIRE please place in the collection box provided. Today if

(vertical text) b.   What time did you get

(vertical text) home today to come

(vertical text) am/pm*

(vertical text) *delete whichever does not app

(right margin vertical text) 7806 * SUTTON HIGH STREET STUDY

(right margin vertical text) USE ONLY FOR STOPS A, B, E & G.

(right margin vertical text) * BUS DEPARTURES SURVEY

7826 THE KINGSTON STUDY	TRANSPORT & ENVIRONMENT STUDIES

INTERVIEW CHARACTERISTICS

REFUSALS   BUS STOP LETTER

DATE CODE   Tuesday 12 December   1   Saturday 16 December   2   Ot⌐

TIME CODE

09.30-10.30	1	13.30-14.30	5
10.30-11.30	2	14.30-15.30	6
11.30-12.30	3	15.30-16.30	7
12.30-13.30	4	16.30-17.30	8

GOOD MORNING (AFTERNOON). WE ARE CARRYING OUT A SURVEY FOR LONDON TR.
MINUTES. IF YOUR BUS COMES WE CAN STOP IMMEDIATELY.

1   Do you work in Kingston Town Centre?   YES   1

2   What is the main purpose of your <u>present</u> journey?   WORK
                                               SHOPPING
                                               EDUCATION
                                               SOCIAL
                                               PERSONAL
                                               ENTERTAIN *
                                               OTHER (Sp

5. STILL THINKING GENERALLY ABOUT WALKING IN BIRMINGHAM CITY CENTRE, HOW DO YOU AS A PEDESTRIAN FEEL ABOUT WALKING IN TH⌐
FOLLOWING PLACES? PLEASE PUT A MARK ON THE LINE (LIKE SO ————✔————) AT THE POINT WHICH BEST
REPRESENTS YOUR OPINION OF THE PLACE, BETWEEN THE EXTREMES OF 'VERY PLEASANT FOR PEDESTRIANS' AND 'VERY UNPLEASANT ⌐
PEDESTRIANS'. IN THE LAST COLUMN, SAY HOW FAMILIAR YOU ARE WITH THE PLACE BY PUTTING A TICK IN THE APPROPRIATE

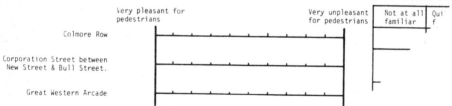

	Very pleasant for pedestrians	Very unpleasant for pedestrians	Not at all familiar	Qui f
Colmore Row				
Corporation Street between New Street & Bull Street.				
Great Western Arcade				

6.01
Most people have a viewpoint about pedestrianisation. It differs
considerably accoding to their interest, and some viewpoints have
been recorded and published while others (the reasons for politically
initiating a scheme, attitudes of the police, say) are much less often
on record. Broadly the viewpoints might be allocated to:

- inhabitants within or adjacent to a   )
  precinct   )   any of these might
- inhabitants of the administrative area)   constitute the
  in which the precinct lies   )   'user group'
- visitors from elsewhere   )
- public transport operators and crew, other transport
  interest groups
- traders within and outside the precinct
- police and other emergency services

All of these people might be classified in another way: are they
giving their views before the event, or afterwards? In other words,
were their views sought as part of a public participation exercise
or have they been extracted from people caught up in later interview surveys, or more complex reviews of a precinct's performance?
Or are the views unsolicited, as part of a campaign for action, such
as the lobbying by major stores for the pedestrianisation of parts
of Central Manchester? (Glasspoole 1980). The emphasis in this
Chapter will be on attitudinal surveys, post pedestrianisation,
taking each precinct on which TEST has information in turn; in some
cases there are a variety of interest groups' responses; other
studies look at several precincts. Traders' views tend to overlap
between this Chapter and Chapter 7.

Surveys of many places

6.02
A survey with wide geographical content was undertaken by OECD (1978)
in conjunction with Europa Nostra, on the precincts of 105 member
cities. There is a mass of data in the Working Document but unfortu-
nately this remains classified as 'restricted' and only a selection
of results can be quoted. In the 105 cities were 130 individual zones
(either a 'unit precinct' or a 'network precinct' to use the jargon
of this book - in other words, independent of other such precincts);
96 cities had only one zone. 32% of the zones were created before

1970 and 62% between 1970 and 1975. There was a marked decrease in air pollution in 72% of the zones, though in only 58% of the UK cities in the sample; pedestrian flows increased in 67% of the zones (one wonders whether many of these were frustrated previously by the lack of appropriate facilities, in much the same way as new urban highways rapidly fill up with apparently newly generated traffic, which was possibly frustrated by congestion). Of relevance to paragraph 6.01 above, 71% of cities invited public response to the exclusion of cars and in 85% of cities shopkeepers and landowners were invited to take part in the decision-making process. One or two other results from the OECD-Europa Nostra study may be found elsewhere in this book.

6.03
Stewart et al (1979) obtained user responses to pedestrianised shopping streets in Birmingham, Bolton, Bristol and Hereford. Responses were sought on personal characteristics, single word attitudes toward evaluating the design of the pedestrian street where the interview took place, and a Likert Scale assessment of environmental change. The report is too elaborate for selective reproduction of its findings. Garton (1977) investigated user responses to five streets in Barnsley and Clyde (1976) was among several researchers following a similar course of study in Liverpool, differentiating between the responses to a fully pedestrianised street and one where buses were also allowed.

Individual surveys

6.04
Turning to attitude surveys in individual precincts, Opinion Research Centre (1976) studied attitudes toward the Saturday pedestrianisation of Deptford High Street in SE London. Bishop (1975) reports a survey on the experimental closure of Commercial Road in Portsmouth in 1972: having a precinct at all received 93% support on a Thursday and slightly less on a Saturday; whether it should carry buses or not was ndt very clear. Bus operators estimated that the precinct would cost them about £57 000 per annum extra due to increased running and operating costs; in Sutton, London Transport assessed their extra operating costs at £30-40 000 in 1979 (TEST 1980). Attitudes of traders and the public toward the Wakefield precinct were elicited in 1976 (Davis, Ives Associates Ltd 1977). 65% of respondents liked the precinct and roughly half were satisfied with access to and adequacy of car parking (the survey is a little obsessive about this). 8 out of 12 traders claimed their trade had been adversely affected, in a range of between 5 and 50% less turnover, the majority claiming about 10 to 20%. These results are decidedly untypical as will be seen from Chapter 7.

6.05
In Chichester (West Sussex CC 1976) the bus operator (Southdown Motor Services) has commented on improved reliability and timekeeping as a result of pedestrianisation. Some of their passengers were also interviewed; those entering or leaving the City via western and southern

routes have benefited much more than those via east and north, though
all passengers have gained. In Bournemouth, shopkeepers were originally against the scheme but now favour it. Some would prefer
buses to be removed from the two bus and pedestrian streets. In
Durham a mid-70s street interview survey found that 50% most liked
the City Centre because of its historic environment, and 35% most
liked it because of pedestrianisation. In a survey in Newcastle upon
Tyne 73% of respondents wanted more pedestrian priority areas (City
of Newcastle 1979). In St John's Road, London Borough of Wandsworth
(1977) 84% of about 570 households thought conditions for pedestrians
had improved since the experiment began (this was at a time when the
road was used primarily by buses and pedestrians but no 'cosmetic
treatment' had been applied).

6.06
Certain oversea responses might be quoted. The GLC (1972) study tour
noted that pedestrianisation of Strøget in Copenhagen was 'strongly
favoured' and that 66% of people interviewed were in favour of the
Gothenburg scheme. In Besançon in France shopkeepers were enthusiastic about the proposals and asked for one street to be pedestrianised ahead of schedule. Pedestrians were asked whether a ban on
buses would encourage them to shop on Minneapolis' Nicollet Mall
(a transit mall, or one sharing the space between pedestrians and
buses). 28% thought it would, 48% saw no change, and 22% thought it
would discourage them. In Uppsala, Sweden, the total traffic volume
was considered unacceptable by between 84% and 65% of residents of
two central city streets: this survey pre-dated the improvements
for pedestrians. Immediately after implementation 60% found the
circumstances improved and 58% of those interviewed favoured more
bus and pedestrian streets (in paragraphs 6.05 and 6.06 the unref-
erenced cities' information derives from TEST 1981).

County Durham

6.07
Further survey results were obtained as a result of requesting information from the UK's Chief Planning Officers. Several sets of results
came from County Durham. In Seaham a postal survey obtained a 47%
(356) response. Of these, 88% found conditions safer, 60% more pleasant, 44% more convenient, than before pedestrianisation. There was
a range of opinion on ease of access by car, which probably explains
why 80% of those travelling to the centre by car favoured making the
scheme permanent, against 93% of bus users and 94% of walkers (Durham
CC 1973). In Crook, looking toward a future when an experimental
scheme was made permanent, 72% of 416 respondents wanted Hope Street
completely closed to all traffic for part of the day (Durham CC 1974).
In Stanley, 300 completed questionnaires were received from a one in
ten households' circulation. Of these, 84% considered shopping to be
safer, 39% more convenient and 62% more pleasant, since the scheme

started. Bus users thought the new bus stop arrangements were better, primarily because of a new bus station. Of those who walked, 27% thought access was more difficult, mainly because of danger in crossing main and service roads to reach the precinct.

Greater London

6.08
In the London Borough of Lewisham (1977) an interview survey was carried out in Catford on its local plan, since the new enclosed precinct was opened. In many ways this survey was about choice between Catford and one of London's largest traditional shopping centres, Lewisham, only about 1.5 km to the north. One of the reasons for lower use of the new than of the old Catford centre as a whole was that Sainsbury's was no longer represented - in other words the quality of the shopping outweighed the environmental, comfort and safety considerations of a new enclosed centre for these people.

6.09
The Carnaby Street Study (Myatt 1975) showed, in its pedestrian interview survey, a 95-98% response in favour of having closed the three streets of the precinct (it is likely that the survey had a greater representation of young people, and of foreign tourists, than we have met in the other precincts discussed above). For nearby Oxford Street Parker and Eburah (1973) discuss survey findings there. 'The great majority' (of 1534 pedestrians interviewed) 'approved of the scheme.' All hotels in the vicinity were approached and their managers thought the scheme had had little effect on them except to exacerbate existing parking problems. Even those residents in the area who responded thought (85%) that the scheme was a success.

6.10
An interesting survey was carried out in Heath Street, Hampstead by the Greater London Council (1975). This street subsided in part in 1967 and was closed for one week in November 1974 for resurfacing. Pedestrians were interviewed on the Saturday of this week (282 of whom 30% were interviewed in the traffic-free area). 62% favoured some form of pedestrianisation of the street. 78% welcomed the diversion of heavy vehicles and 67% said they were disturbed by the general level of traffic in the area. 88 local business managers were also interviewed. Overall, nearly half of the managers and three fifths of the pedestrians liked the conditions created by temporary removal of traffic (despite the inconvenience of the roadworks and without any of the environmental pleasantries we associate with full pedestrian schemes) and were favourably inclined toward some form of pedestrianisation; four out of five businesses had been satisfied with the temporary servicing arrangements. Heath Street, Hampstead, a narrow shopping street on a hill, remains, in 1981, heavily trafficked.

6.11
Humberside CC (n.d.) surveyed pedestrians in three of their centres
and compared the results with the Ossett, West Yorkshire, precinct.
The summary of results is reproduced below, the table showing a high
level of approval, wide variations in the modal splits, and wide
variations in adequacy of the car parking provision:

	USAGE			ENVIRONMENTAL IMPORTANCE				MODE OF TRANSPORT				PARKING FACILITIES			
	No. of shoppers	More	Less	Environment improved	Safer	Easier movement	Quieter	More pict-uresque	Car	Bus	Foot	Train	Adequate	Not adequate	Too few spaces	Remoteness from shops
Ossett Town Centre	80	30	5	95	70	80	60	70	25	15	60	0	68	31	20	80
Whitefriargate Hull	75	47	1	94	76	56	39	40	38	48	10	4	25	75	83	32
King Street Bridlington	88	20	3	70	58	30	18	5	45	17	32	1	17	83	82	17
Toll Gavel Beverley	88	10	11	-	-	-	-	-	57	12	26	0	-	-	-	-

*VI.01 : Summary of pedestrian interviews in Humberside towns, using
Ossett as a control. All figures are percentages of the relevant
sample studies, eg only car drivers answered parking questions.*

6.12
Pembroke Dock had an experimental scheme that was later removed as a
result of a local council decision (South Pembrokeshire DC 1980) which
did not accept its officers' recommendation to retain the scheme.
However, the decision 'has left a sympathy for pedestrianisation or
at least more severe car restraint. The second improvement could
well be effected at Tenby.' In Pontypridd (Mid Glamorgan CC 1980)
the County observe:

> 'As far as Pontypridd is concerned, the greatly reduced number
> of vehicles using Taff Street has caused pedestrians to become
> more careless of other road users and demand more priority in
> crossing the street. The perceived speed of buses seems to be
> far higher now that they are the only vehicles making a deli-
> berate through passage of the shopping centre. Bus drivers
> have complained of the pedestrian attitude.' The County go on
> to say 'Shopkeepers' attitudes are always negative towards any
> change in their operating environment and several partial

pedestrianisation schemes have been abandoned due to
pressures of this sort.'

In South Glamorgan (1980) the County's view is that, despite economic
difficulties in the area (loss of a steelworks, for example) traders
in Queen Street, Cardiff have noted that the level of trade has not
fallen. They explain this through the greater attractiveness of the
precinct since much traffic was removed.

6.13
In Scotland, the City of Dundee (1976) has carefully monitored its
pedestrianisation. 2300 were in favour of retaining the pedestrianised Murraygate compared with 115 against. There should be improvements to landscape and lighting, the people interviewed suggested,
and the scheme should be extended to the High Street. Furthermore,
some toilets should be provided. There was concern about insufficient
parking in the vicinity of Murraygate.

6.14
The Borough of Thamesdown (1980) surveyed shoppers using Swindon town
centre in 1978. 'In the light of previous criticism of the relative
concentration of bus routes and stops at Fleming Way and the Town
Hall, a question was included which asked how convenient people found
this distribution.' From the 500 shoppers interviewed 73% found it
fairly or very convenient. Norfolk CC (1974) appraised the costs and
benefits of pedestrianising London Street. 290 interviews in 1967
overwhelmingly favoured the scheme that had just been introduced, and
a follow-up survey in June 1971 produced 100% in favour. 55% of
those interviewed said they frequently walked through the street
even if they had no specific reason for visiting it. This County
also conducted interviews in King's Lynn (Norfolk CC 1975). 97%
(of 380 people) considered that the area had improved since it had
been closed to traffic. There was a negative response toward allowing buses in the pedestrianised street.

6.15
On balance, the users of pedestrian areas are enthusiastic about them.
They criticise points of detail (particularly about parking facilities)
and are happy to be consulted before pedestrianisation is introduced,
though this happened too infrequently in the early days of street conversion in Britain; furthermore, most of the survey results reported
here are after-the-event ones. Shopkeepers generally favour precincts,
more often when they are well located within them than if they, the
traders, are outside the precinct.

7

making or breaking

*If there has not been a positive effect on trade (or at least it has
not deteriorated) then no precinct can be considered wholly successful. The strong antipathy held by shopkeepers toward pedestrian
schemes often delayed, sometimes aborted, early projects. However,
within the last 5-10 years traders have begun to realise that most
precincts, worldwide, are accompanied by turnover improvements; in
the early-1980s there are many instances where shopkeepers and other
businesses are demanding that their street should become a precinct.*

*As with Chapter 6, information can be split into many-place and single-
place studies. A large number of these are discussed, and the success
stories do seem far to outweigh those where trade has actually diminished. Indeed, some of the latter's difficulties can be attributed
not so much to the precinct but to consumer preferences for shopping
centres elsewhere or to mail ordering. Very occasionally a shopkeeper
will suffer because the routeing of pedestrians has changed so they
no longer flow past his door.*

7.01
This Chapter is concerned with one of the most important measures of
the success of precincts: their economic impact. It will look at this
mainly through the effects on shopping turnover but will also note
land use, shop type and floorspace changes, mainly within town centres.
It will not neglect those who claim to be losers from the complex
ideology called pedestrianisation, which necessarily introduces conflict.

7.02
Changes in trading patterns are of course not simply the results of
pedestrianisation. Shopping turnover depends on what is being sold,
by whom, in what kind of sales environment, as part of what external
economic condition, under what type of media persuasion, and so forth.
Another profound influence on traditional shopping streets has been
the advent of out of town centres, hypermarkets, cash and carry stores.
The growth in mail order selling has also been important. So, the
catchment areas of traditional stores has changed; small shopkeepers
have been hit by the trend toward largeness and car-oriented central-
isation. (The Brent Cross Study (Lee and Kent 1977) looks at many
explanations of the success of such centres. Major reasons for shopping
there were parking facilities, one stop shopping or layout, and quality/
choice. Car ownership of Brent Cross users compared with Greater London
was: no car 13%/50%; one car 51%/42%; two or more cars 36%/8%, respectively. The Caerphilly Hypermarket Study (Lee and Kent 1975) and
Schiller (1981) also help us to understand trends away from central
areas - though it is worth noting that higher energy prices are in
some cases beginning to reverse these trends). Many shopkeepers,
having learnt that pedestrianisation in general improves turnover, now
advocate the creation of precincts to stem the outward drift from central area shopping: this has been commented on earlier in this book.
In order to keep central areas both economically viable and acting as
the prestigious centre that towns and cities are deemed to require,
many city fathers have included pedestrianisation in their armoury
of resilience among changing circumstances.

7.03
Most of these largely economic effects have been prevalent during the
1970s, the peak period - as we have seen - for the introduction of
precincts in Britain. Morrell (1980) has shown that the growth in
services 1973-78 dropped to 2% per annum from 3.4% per annum in the
period 1964-73 and that retail prices rose by 16.7% per annum in the
later quinquennium by comparison with 5.9% per annum in the earlier
one. Spending on food, according to Morrell, has been sensitive to

extreme price movements, the rapid increases of the mid-1970s resulting in cut-backs in spending in volume terms. Perhaps the 1970s peaking in the creation of precincts resulted from planning in the 1960s (Sutton's High Street was first planned to be pedestrianised in the early 1960s: the job is still unfinished - TEST (1980)); perhaps the creation of precincts has now gathered such a momentum that these price fluctuations do not materially affect their future.

7.04
Sometimes pedestrianisation can be too successful. 'London's most fashionable specialist shopping streets are being killed by their own popularity' according to Sudjic (1980). Reading further into his article, however, suggests that the murderers are avaricious property owners rather than popularity. Discussing London's South Molton Street, he quotes rent rises from £145 a week to £1250 in one case, a 625% rise in another, and an annual rise in one step from £40 000 to £160 000 in another.

7.05
To conclude this introductory section it is useful to look again at the mode of travel/turnover survey in the West Midlands (West Midlands PTE 1979). At November 1977 prices the report suggests that car-borne shoppers in strategic centres (Birmingham, Coventry, Wolverhampton and six non-regional ones) spent an average of £12.55 on each shopping trip compared to £6.81 for bus-borne shoppers. Then the report argues that for strategic centres bus contributes 41% of *retail turnover* compared with 43% by car, but in Birmingham City Centre (where 65% of shoppers use public transport compared to 19% using cars *to reach* the Centre) the public transport contribution to retail turnover is 56%, compared with 33% by car-borne shoppers. The Birmingham findings, however, are unique in the West Midlands: 'car makes a much more significant contribution to turnover in Solihull, Sutton Coldfield and Stourbridge' and the same can be said for the smaller District Centres.

7.06
The report also discusses retailers' attitudes, which show strong feelings for more car parking in the City Centre, and also improved bus penetration. Overall, it seems, retailers did not want more pedestrianisation, though these feelings were stronger in Coventry than they were in Birmingham. The West Midland PTE's findings seem to have confounded many people who thought the car was paramount in shopping trips (that this is not universal has already been pointed out: people who walk all the way to and from the shops are often a highly significant proportion of the modal split. In Birmingham's case it has been shown that if the public transport service is good enough - in terms of convenience, fare levels, choice of routes, etc - it will attract the largest proportion of shoppers travelling to a major centre). It seems that Birmingham's experience must be taken very seriously indeed by transport, and shopping precinct, planners.

82

7.07
One document which will be in many planning offices is the Civic
Trust (1976) short paper on the effects of pedestrianisation on trade.
The results can be tabulated, and are a collation from many references
already used in this book: Arrive (1971)[1], Wood (1969)[2], Jennings et
al (1972)[3], GLC (1972)[4], OECD (1974)[5], Institute of Traffic Engineers
(1966)[6], Municipal Journal (1975)[7] and Myatt (1975)[8]. The table below
shows this material; the superscripts identify the source.

City	Change	Source
Atchison, Kansas	18% increase	6
Cologne	25-35% increase	4
Copenhagen	25-40% increase	4
Durham	20% reduction: traders asked for buses to be reintroduced in city centre	7
Dusseldorf	36-40% increase	4
Essen	25-35% increase after initial decline	4
Gothenburg	Range: 20% reduction to 10% increase	5
Hamburg	70% of shopkeepers noted increase in sales	5
Hereford	10-15% increase; one case 25-50%	1
Kalamazoo Michigan	15% increase	6
London, Carnaby St	Is pedestrianisation a good idea? 81% of shopkeepers said 'yes'	8
Minneapolis	14% increase	4
Munich	about 40% increase	4
Norwich	of the 32 shops in London Street, 30 showed effect on trade in first 6 mths 28 increased trade	2
Pomona California	16% increase	6
Rouen	10-15% increase	5
Vienna	20% increases noted by 60% of mer-chants	4
Watford	72% of shopkeepers said pedestrianisa-tion had favourable effect on trade	3

VII.01 : Changes in shopping turnover after pedestrianisation

7.08
Any trader doubting the success of pedestrianisation must be reassured
by the above table. If they need more evidence, OECD (1978) and much
of the remaining material in this Chapter must help. The OECD-Europa
Nostra study, as has already been said, elicited responses from
105 of its member nations' cities about their pedestrian precincts.
The distribution of the cities is as follows: 23 in Austria, 19 in

Germany, 1 in each of Greece, Finland, Ireland and Portugal; 6 in
Italy, 4 in each of Netherlands, Switzerland, Denmark and France;
13 in UK, 3 in Norway, 2 in each of Sweden and Australia; 5 in each
of New Zealand and Canada, 6 in USA. For these cities the following
information on retail sales is quoted from the report. It should
again be borne in mind that the data have been obtained from a pro-
visional OECD Working Document:

> '49% of all pedestrian zones recorded upward trends in
> turnover rates, particularly in Austria, Germany and Scan-
> dinavia (71%, 63%, 67%). The range of increase lies mainly
> around 25%, with occasional upward thrusts as far as 50%.
> 25% of all pedestrian zones have not registered any marked
> sales improvement, particularly in the UK (50%), North
> American Cities (46%) and Pacific Country Cities (67%).
> 2%, however, report a reduction of turnover rates.'

7.09
OECD go on to ask whether a successful pedestrian zone might not im-
pinge negatively on the image and hence the business of adjacent non-
pedestrianised areas. They say this is not inevitable and that in
21% of all cases a positive effect on neighbouring streets was indicated, against 13% reporting a deterioration, especially in Germany.
Another fear when a precinct becomes successful is whether shops and
their merchandise will deteriorate. OECD discovered that this happened in 18% of the precincts, but did not do so in 77% of them.

Studies of individual places

7.10
As in Chapter 6, Humberside CC (n.d.) provide us, Illustr. *VII.02*,
with a table in which precincts in Bridlington and Hull are compared with Ossett, West Yorkshire. This shows the number of units
by type of business, loading problems, effect of pedestrianisation
on trade, and parking characteristics. Bridlington has a slightly
higher proportion thinking that trade is worse than those thinking
it is better, around a core of 66% perceiving no change. Both Hull's
and Ossett's 'better' voters were a higher proportion than those
saying 'worse'.

7.11
The London Borough of Enfield (1980) provided information on the
Edmonton Green Shopping Precinct. 157 shop units were demolished and
replaced by 140 units, the total floor area on two levels, in the new
precinct, being approximately 28 000 m^2. The second stage of the
Centre was completed in 1973, and it has 792 car parking spaces and a
bus interchange at the main entrance. 'The Centre appears to be trading
successfully and...it is a safer and more pleasant environment in
which to shop, when compared with the open shopping streets astride
a major road that it replaced.'

TYPE OF BUSINESS					LOADING		AFFECT ON TRADE					PARKING			HABITUAL PKG. AREA		
Supermarkets	Shops	Offices	Banks	Others	Problem	Aggravated by pedestrianisation	Better	Same	Worse	Best market day	Other	Have employees travelling by car	Difficult to get all day parking	Multistorey	Surface c.p.	Private c.p.	On street parking
17	83	-	-	-	25	25	25	59	16	-	-	100	16	-	25	75	-
4	63	23	6	4	50	50	26	74	0	-	-	94	37	37	25	24	15
11	55	7	11	15	48	44	15	66	19	30	7	62	30	-	15	33	15

VII.02 : Trader interviews in Humberside and West Yorkshire
I = Ossett Town Centre (an 'ideal' control site outside
Humberside); II = Whitefriargate, Hull; III = Kings
Street, Bridlington
All figures are percentages of the relevant sample studies

7.12
The City of Lincoln (1980) pedestrianised their High Street and 'the
local Chamber of Commerce is of the opinion that the volume of trade
in (it) has increased significantly since implementation but how far
this can be directly attributed to it rather than to an overall trend
is hard to say...with the establishment of peripheral shopping centres
aimed mainly at the car-borne shopper and making available both convenience and durable goods, then pedestrian schemes of this sort
become an important element in enabling city centres to compete...a
good indication of the economic effects of the scheme in Lincoln is
the attitude of local traders who have actively supported extensions
to the original scheme, there is obviously some feeling of not wanting
to be left out and this would indicate a positive economic benefit in
being located in the paved areas.'

7.13
Gwynedd County Council (1980) revealingly say 'the only scheme operating at present is in the shopping centre of Bangor...shop owners are
generally in agreement with the scheme. Whilst this has certainly
made shopping safer for pedestrians there have been newspaper comments
about the money for such schemes being better spent elsewhere (although
there is very little spent on pedestrianisation in the County)'.
Chichester's elaborate report on its pedestrianisation (West Sussex CC
1976) has a substantial section on trader responses. 55% of those
inside the precinct favoured it while only 38% of those traders outside

the precinct felt that way. Those inside the precinct who felt trade
had improved balanced those who felt it had worsened (around 56% who
detected no change or did not respond), while of those outside the
precinct 8.2% noted an improvement, 14.5% a deterioration, and 77.3%
no change or no response. There is much more of interest on trading
changes in Chichester that cannot be explored here except to quote
the report's comment on traders' attitudes to convenience of access
for people and goods: 'Very few traders, it seems, consider conditions
are improved and a substantial number consider matters have worsened.
In so far as public transport is concerned, this latter opinion is at
variance with the opinion expressed by the bus operator and passengers.'
Could it be that the traders' concern in other areas is a little over-
played?

7.14
Durham County provided trader information on three locations, and
further information was collected for the Durham City case study reported in TEST (1981). Stanley Centre is reported in Durham CC (1975):
37 traders directly affected by pedestrianisation contained 5 experiencing no change, 10 an increase, 15 a decrease. In Crook (Durham CC
1974) there were 35 responses from traders: 15 no change, 10 an increase, 10 a decrease in trade. Nevertheless 80% wanted the scheme
to be made permanent, and represented 47% of the total number of
traders in the street. Seaham (Durham CC 1973)provided 54 responses
from traders of whom a smaller set responded to turnover changes: 13
no change, 10 an increase, none a decrease. In Durham City the picture
is much brighter: 100% of traders approached thought the scheme was
generally successful and, for the period June 1975 to end 1979, 3%
thought trade had decreased, 44% that it had remained the same, and 52%
that it had increased. There has also been a substantial increase in
shopping floorspace - the County Structure Plan initially tended to
limit this to an extra 10 000 ft^2, though this has since been reconsidered: the 10 000 sq ft has been built, approval has been given to
a further 55 000 ft^2 and applications for another 16 000 ft^2 were
being considered in 1980 (the metric equivalents are respectively
929, 5109 and 1486 m^2).

7.15
Pedestrianisation is an economic success. A majority of a large number
of traders asked about changes in turnover after a precinct was created
believe it had increased. (There is little published data on newly-
built enclosed shopping centres, but the Brent Cross Centre and Caerphilly hypermarket have both done well). In addition, shops change
their image, improve their shopfronts, and enlarge their floorspace,
all of which are indications of success.

8
futures

The future is very uncertain, though the idea of shopping precincts is now far too strongly held for their creation to cease: probably the rate of increase will decline, for much has already been achieved.

However there remain many places with no proper facilities for walking to and around the shops at all. Other places could seemingly do much more than they are doing. London is one such place and its good intentions are compared with a low level of achievement, together with some explanations for non-events.

Clearly the pedestrianisation of certain shopping streets is extremely difficult to achieve - their traffic loads (and often many bus routes) are real enough to require alternative routeing and this can be very expensive. A new look at such streets is required, where all users get a more equitable share of the movement space.

Many authorities' future proposals are reviewed.

8.01

If the rate of growth in creating pedestrian shopping precincts were
to be an extension of the UK curve in Illustr. *I.08* then all of Britain's
shopping streets would have been converted to precincts, and much of
our cities' central areas would have been taken over by enclosed shopping centres, before the year 2000. To continue the curve must be
wrong, for a number of reasons, foremost among which is a shortage of
public sector money. Land shortages may curtail activity, when coupled
with conservation policies. Town planning may be swayed by rural settle-
ment or outer urban needs against inner city ones. Another reason could
simply be disinterest on the part of local politicians and officers, or
a redistribution of resources away from town or suburban centre renewal
toward other urban priorities.

8.02

Certainly, from the perspective of 1981, the peak in the conversion of
all purpose streets to pedestrian precincts seems to have passed. The
building of new enclosed centres may be rather like the office-building
boom of the last two decades and continue until we are all surfeited
with shopping facilities. However, this rather forbidding vision lost
credibility recently: Building Design (1981) carries an article explain-
ing that Thamesmead's long-awaited shopping centre has been abandoned
by Costain (the developer) on the grounds that 'it is not viable in
the current market conditions.' Despondency apart, the civilising of
existing shopping streets remains a problem in many localities and
we can expect improvements to continue, if at a slower rate than during
the 1970s. Paragraph 8.06 below, in making a plea for a more equitable
distribution of movement space between the users of a street where full-
scale pedestrianisation would be too costly, and paragraphs 9.06-9.08's
plea for a different kind of urban sanity, and the second and third planks
in this concept of 'affordable Utopias'. How well is London meeting
these challenges?

London: the art of the possible

8.03

Much of London is blighted by what appears to be a greater emphasis on
the movement of vehicular traffic than on the movement of people on foot.
London has innumerable linear shopping streets, a large number of which
are classified as Metropolitan or as Trunk Roads. The GLC and the Department of Transport are respectively responsible. Their powers override

those of the London Boroughs in whose areas the shopping streets lie.
There do seem to be difficulties in coordinating different interests,
rather than a lack of will from any of the individual interests. Doubtless there are excellent reasons, but the GLC has reviewed many pedestrianisation schemes many of which, over the years, seem unaccountably
to have disappeared from sight. Thus, a Press Release (GLC 1974) lists
24 shopping streets where a car ban was proposed. Those with a full
7-day closure were in Pinner, Wood Green, Tottenham, Havering, Woodford
Green, Brixton, Southwark, Battersea, Kensington, Hammersmith (this last
one was Monday to Saturday market hours only). Romney Road Greenwich
was to have been bus and pedestrian only at weekends, peak occasions
for visits to this historic area. 13 other streets were designated for
Saturday closure only.

8.04
A progress report in 1977 (GLC 1977) showed that of the 11 proposed
7-day closures, only those in Wood Green and Tottenham were operating.
Romney Road was still being discussed in 1977. The reasons for abandon-
ment of the others are:

Pinner High Street	Members, the Borough and local residents all disliked the scheme
Farnham Road	Rejected by the GLC's North East Area Board against officers' recommenda- tions
Mill Lane	Approved by NEAB but rejected by Borough Committee against Borough officers' recommendations
Brixton Market	Borough abandoned due to financial difficulties
East Street	Borough and traders objections
Battersea High Street	Full scheme abandoned because of Borough and traders objections but Saturdays-only closure likely
Beauchamp Place	Rejected by Area Board as effects of traffic diversions would be unaccep- table (presumably by residents in local streets)
North End Road	Diverted traffic could not be accom- modated in other streets

8.05
There is also GLC documentation on the futures of pedestrianisation in
Blackheath Village, Morden and Orpington Town Centres, South Kensington
Station Area, High Street North, East Ham, and doubtless other places.
Each is characterised by clashes of interest groups such that one wonders whether anything on the scale of provincial towns and cities' pedestrianisation will be achieved in London. The list in paragraph
8.04 reinforces this disbelief. London possesses many streets not
even mentioned so far that, from the pedestrian's and trader's point of
view at least (to say nothing of the tourism and civic pride interests),
cry out for pedestrianisation; at the very least this should cover peak
shopping times like late closing days - Thursdays and Fridays often -
and Saturdays. In London's international centre there are streets of
major importance like Regent and Bond Streets, Piccadilly, Knightsbridge,

the Strand, Charing Cross Road, Shaftesbury Avenue, Tottenham Court Road.....

8.06
The argument against these closures will be that they are principal traffic routes. Does this matter on two evenings and Saturdays? And how has Manchester managed to close streets with the strategic regional importance of Market Street and Deansgate? (Glasspoole 1980 and Greater Manchester Council 1981). Because motor traffic so dominates most people's lives, perhaps some of the arteries which seem difficult to pedestrianise (primarily because the scheme cost is far more the result of providing alternative traffic routes than it is with the repaving, lighting, planting and other cosmetic treatment of the street itself) should be urgently reappraised as to the share of movement space allocated to each means of transport. The difficult streets - the London ones mentioned plus a reconsidered Oxford Street and, for example, New Street and Corporation Street in Birmingham - should give all users a share according to criteria such as efficiency of use of movement space, energy efficiency, number of people moved / m^2 / second or..... more abstract concepts like civility, gracefulness, serendipity, fun. The pedestrian would then not only receive more space, but more frequent, safer, convenient crossing facilities. In fact there is a whiff of this concept in the latest of a long line of proposals for London's Piccadilly Circus (GLC 1980b): walkers could have a greater share of the total movement space than they have now.

Realities

8.07
OECD's and Europa Nostra's survey of 105 cities (OECD 1978) considers the future of pedestrianisation. They say 'Although the movement to create pedestrian zones in city centres has reached its culmination point because most cities already have one and see no possibility for creating more, most zones (80%) reported that the existing zone would be expanded. Only 17% said they did not have such intentions.' They go on to say 'In order to maintain the commercial structure of the downtown area or to reduce overall traffic volume, 58% of all cities have a firm strategy not to increase the number of major shopping centres on the urban fringe. The average, however, hides two different trends - that of Western European countries which have adopted this strategy (74%) and that of Scandinavian, North American and Pacific countries where only 20%, 9%, and 14% respectively favour this strategy. For 53% of cities supporting this strategy the main reason is the protection of the existing commercial structure; 15% do it in order to reduce traffic.'

8.08
Various planning documents help an appreciation both of the importance of shopping precincts, and their likely introduction in the future. The Greater London Development Plan (GLC 1976) is encouraging at first sight. Within five paragraphs devoted to pedestrians perhaps the key statement is 'Wherever possible provision should be made for the pedestrian away from the noise and fumes of the traffic, enabling him to move safely in a more pleasant environment. This should be one of the

most important objectives...to be achieved in large-scale redevelop-
ments and in environmental improvement schemes.' It may be restrictive
that the sanity, soundness and safety of the walker do not of themselves
determine provision, but come along as part of a package of objectives.
The 1980-1985 GLC Transport Policies and Programme (GLC 1979b) has one
solitary sentence, in 130 pages, on pedestrians: 'Pedestrian Measures -
schemes designed to improve pedestrian safety - will continue to be
introduced.' There is considerably more about pedestrians in the
Dorset TPP (Dorset CC 1978) - a County with about one fourteenth of
Greater London's population..

8.09
TEST (1981) shows, in its case studies, some towns' proposals for the
future. Barnsley planned removal of buses from two of its pedestrian
streets in 1976, then deferred this, then effectively did it while
repaving of the streets took place in 1980, though on an experimental
Order: the bus operator may object to this. In Birmingham, as already noted, the pedestrianisation of New Street and Corporation
Street would be quite difficult, but is still desired; the PTE is
strongly opposed to removal of the many buses using these streets.
Nevertheless the City feels it would have partly pedestrianised these
two streets by now had there not been local government reorganisation.
Bournemouth has long-term plans for moderate extensions to its pedestrian streets. In Durham City there is little shopping street left
to pedestrianise so they appear content to introduce Orders to restrict
loading on one street, and access on another, and to improve the environment of North Road. Exeter proposes an extension to take in the
remaining all-traffic part of the High Street - again the network is
so extensive there is not much else to do. In Newcastle-upon-Tyne
public pressure is mounting toward the removal of buses from Northumberland Street; in 1981-82 the City intends to add Blackett Street
to its pedestrian network. So strong is the support for pedestrianisation that other schemes may be progressed, particularly if the three
new Metro stations bring even more shoppers into the Centre.

Other pointers to the future

8.10
Several Districts in Wales mentioned their plans for the future.
Llanelli BC (1980) has a long-term objective to pedestrianise *all*
streets within its ring road. Gwynedd CC (1980) propose a scheme in
Caernarfon as part of the Menai Strait Local Plan; attempts to pedestrianise the main shopping street of Holyhead have been frustrated
by the local bus operator. Elsewhere, and often because of hilly
sites, the County have faced difficulties in locating alternative
parking and rear servicing facilities. Another prospect Gwynedd are
considering is the seasonal closure of village centres which are
subject to heavy tourist pressure. Aberdaron, for example, experien-
ces pedestrian-vehicle conflict in peak periods but in this case
no alternative route for through traffic has been found. South
Glamorgan CC (1980) noted that while they already have a good network
in Cardiff, they have authorised six new arcades to be built.

9

is it all worthwhile?

The shopping precincts that have been created form a minute proportion
of the total movement space of any town or city. While these precincts
are very much welcomed, they do underline the inadequate provision elsewhere, and there might be a temptation in some authorities to think
'Well, we've dealt with the pedestrian, what problem shall we tackle
next?'

People walk with a mixture of purposefulness and randomness; the latter
characteristic tends to be banished by traffic engineers who try to
channel people, and induce them to cross roads by subway. Many walkers
are injured or killed while revolting against these unnatural restraints.

The book's main text concludes with the view that far too much urban
movement space is allocated to motor vehicles and far too little to
people walking. With sentiments that are close to being heretical, it
finally suggests that much of our road space should be turned over to
activities where people can actually enjoy themselves.

9.01

Most people would agree that improvements in conditions for pedestrians
are worthwhile. Anything that reduces the risk of death or disability,
impaired hearing, inhalation of noxious fumes...fairly quickly reaches
the Statute Book in other domains of life. If, in addition, the creation
of places strictly for walking uplifts the soul, encourages social encounters, makes shopping pleasant, in other words enhances the quality
of life...then it must be valuable.

9.02

But, what are we discussing? 1304 units that have surfaced in this
study, perhaps 1450 in all in the UK. Many quite large towns do not
have any shopping precincts fit for pedestrians and others, as we have
seen, perform badly in comparison with other towns or cities. The delay
is often attributable to the British obsession (unlike N American and
W German authorities, both of whom are doing very well on pedestrianisation) for off-street servicing and for new roads to take displaced
traffic. Both these devices are very expensive and probably not cost-
effective.

9.03

Of the verifiable 1304 units, the length of 907 is known to be 136.13 km.
Assuming a rough consistency in the other units that lift this to a
likely total of 1450, then the total length would be about 217 km, or
about 215 km in Great Britain. The total public road length in GB is
some 336 000 km (DTp 1980b). Simply to grasp the scale of present-day
provision, the ratio of public multi-purpose road to pedestrian shopping
street is approximately 1563:1.

9.04

There are wide variations in this ratio. Taking only urban roads, for
London it is 1132:1 (GLC 1979b) while for Doncaster it is 765:1 and for
Rotherham 4100:1 (S Yorks CC 1980). What these ratios tell us, albeit
extremely crudely and with many reservations, is that there are vast
areas of towns and cities with, at best, footpaths along each side of
the road, footpaths at times inadequate for those using them, or which
simply evaporate when motors demand more space, footpaths which are
broken at each side road and which connect with the other side of the
main road quite infrequently. And, as pedestrian densities reduce
toward the edges of towns the need for special facilities for those
walking becomes greater. This has been well recognised in Holland. The
'Woonerf' system of suburban road space-sharing between different road
users, allowing less space for cars and more for kids to play for example,
was noted in a visit to Delft by Dalby and Williamson (1977), and a paper
on this topic will be presented to the 1981 Town & Country Planning
Summer School (Jonquiere 1981).

95

9.05
In fact, looking at large numbers of suburban housing layouts you
would imagine that their designers thought that Radburn was a coal-
fired panel heater. It is in these very areas that children and old
people are particularly at risk because they have a sense of false
security: 'there is not much traffic about so it is safe to cross
the road'. The author's son had his leg broken, and a colleague's
girl friend was killed, while both were probably thinking these
thoughts, that is, if they were thinking about such a mundane topic
as traffic at all.

9.06
Cities and towns are not places where every street around each individual block should witness the public mating of squirming vehicles,
their population burgeoning like *Escherichia coli* in the human gut.
If there must be motor vehicles there are three possible places for
them - above or below the walking plane, or concentrated on a moderate number of separate roads ('motorist precincts') that link to
dead-end filaments that allow such servicing of buildings as has to
be close-up. If this were accepted, and having solved the shopping
problem for pedestrians (at least in a large number of places), we
could move on to commercial and office centres, to manufacturing
industry and then outwards to the residential suburbs, in a major
attempt to civilise the areas where most of us live. (We would,
of course, have gratefully absorbed those other sensitive places
for pedestrians - educational and health precincts, churchyards, parks,
gardens, recreation areas - that have thoughtfully been provided.)

9.07
Sir Walter Raleigh (1861-1922, not the Elizabethan sailor) wrote:
'I wish I loved the Human Race, I wish I loved its silly face, I
wish I liked the way it walks....' Apart from being an all-time
nihilist, Raleigh seems to have been one of the founder members of
a traffic engineering institute. For it is *because* people often walk
randomly (note the *Ramblers'* Association) that railings line our
footpaths and attempts are made to push walkers into subways and up
on to bridges: in other words, they are stopped from doing what is
natural, which is to walk freely on the ground. This is often purpose-
ful, but it should be possible to make split-second deviations from
your path.

9.08
In any place unrestrictedly set aside for walking, we are free to be
random, and silly, if we wish. The ultimate achievement would be to
turn over a high proportion of urban movement space to walking, people
in wheelchairs, shops, cafes, roller-skating and roller-coasters, pro-
cessions, bazaars, tattoos, kicking footballs, circuses, Ferris wheels,
perambulators, experimental buildings, sound and light shows, sand
pits, solar collectors, buskers, food growing, sculptures, sheep gra-
zing, exhibitions, trees, expanses of water, Punch and Judy, brass
bands, flower sellers and school crocodiles. How many of these things
happen on the roads of your town?

10
list of UK schemes

Explanation of columns 1-14 for the following tabulation, showing all known pedestrian units in the United Kingdom.

TABLE
COLUMN
NUMBER

1 a. County Councils in England and Wales
   b. Regional Councils in Scotland
   c. States in Guernsey and Jersey
   d. Departments of Housing, Local Government & Planning in Northern Ireland.

2 a. England, Wales and Northern Ireland District, Borough or City Councils
   b. Scotland - District Councils

3 Place - name of town, village etc.

4 a. Population figures (in thousands) for County or equivalent, and District Councils were taken from George Godwin Ltd (1974) and correspond to the period after local government reorganisation
   b. Population figures for places within Districts derive from the 1971 Census
   c. Scottish population figures were taken from the Planning Directory (1975), edited by Teresa Langton.

5 This is the year in which the unit was pedestrianised.

6 A - covered units, also used where subdivision A1 or A2 not known
   A1 - almost entirely pre-1939, primarily arcades
   A2 - purpose-built, post 1945. Usually enclosed shopping centres with one or more 'malls'
   B - uncovered streets, also used when subdivision B1 or B2 not known
   B1 - originally all-purpose shopping streets containing all types of traffic, since converted to streets primarily for walkers
   B2 - purpose-built, uncovered, post-1945
   C - type B streets that also permit buses.

7 Permitted users. These people or vehicles are permitted access to a pedestrianised area, sometimes under special conditions:

   A - Access   C - Cyclists   D - Disabled drivers
   P - Permit holders   S - Service vehicles T - Taxis
   U - Public utilities

8 The length of pedestrianised section is that given by local authorities or scaled from maps

9 This shows the type of access that servicing vehicles have, in

1	2	3	4	5	6	7	8	9	10	11	12	13	14
	LUTON	Lime Street		Early60's	B1		20	1-5				c	3,6
		Old George Yard		1976	B1		90	1-5					4
		Silver Street*	161.2 159.4	1976	C	D,C	106	1/4					1
		LUTON											
		Arndale Centre		1978	A2		630		70				1
		Bute Street		1978	B1		45	5					
		Cheapside		1978	B1		32	5					1
		Solway Road		1978	B1			5					
		South Road		1978	B1								
		Town Square area		1978	B1								
		Williamson Street		1978	B1		40	5					4
	SOUTH BEDFORD-SHIRE		88.3										
		LEIGHTON BUZZARD	20.2										
		High Street*		1977	C		36	1/4					
		Market Place*		1976	B1		13	4					3
BERKSHIRE			620.0										
	BRACKNELL		63.9										
		BRACKNELL NEW TOWN											
		Town Centre	33.7	1950+	B								5
	READING	READING	132.0										
		Broad Street	127.0	1970	C	D,S,A	c.200	2	200	230	1		5,6,9,13
		Butts Shopping Centre			A2		155	5			0		3
		Queen Victoria Street		1970	C	A,S,D	c.50	2			2		
		TILEHURST	98.5										
		The Triangle			B1						1		3
	SLOUGH	BURNHAM											
		Wentworth Avenue			B1								1
		SLOUGH	86.0										
		Covered Centre		1972	A2		490	5	70	230	3		5,6
		High Street		1976	C	D,S,A	580	2/4			2		3,6,7
	WINDSOR & MAIDENHEAD	MAIDENHEAD	122.7										
			44.2										
		High Street (part)*		1974	B1		240	3/4			4		5
		Nicholsons Walk*		1969-74	A2		175	4/5			2		5
		WINDSOR	16.1										
		Church Lane*		1977-78	B1		50	2					
		Church Street*		1977-78	B1		65	2					1
		King Edward Court*		1980	B2			5					1
		Lower Thames Street(pt)*		Early70s	B1		80	1					14

1	2	3	4	5	6	7	8	9	10	11	12	13	14
	WOKINGHAM	Market Street*		1977-78	B1		60	2					1
		Queen Charlotte Street*		1977-78	B1		15	2					1
		Windsor Bridge*		1974-75	B1		70						1
	WOKINGHAM		98.8										
		Crockhamwell Road	76.2										1
BUCKINGHAMSHIRE	AYLESBURY VALE		476.0										
		HADDENHAM	115.1										
		Dragontail*											1
		Turnstile*											5
	MILTON KEYNES	BLETCHLEY	66.8										
		Queensway	30.2	1976	B1/B2		150	4					1
		Duncombe Street*		1978	B1		5	1					4
		MILTON KEYNES	46.0										
		Shopping Centre		1979	A2		700	5					6
		WOLVERTON	13.8										
		Radcliffe Street		1978	A2		50	5					1
CAMBRIDGESHIRE	CAMBRIDGE	CAMBRIDGE	505.0										
		King's Parade	98.5	1975	C	C,S,A,D,T	230						6,7,13
		Lion Yard Centre			A2								3
		St. Andrews & Sidney Sts	88.5	1975	C	C,S,A,D,T	250						3
		Elizabeth Way*											1
		Senate House Hill*									1		1
		St. Anthony Way*									0		1
		St. Mary's Passage*									0		1
		Sussex Street*									0		1
		Trumpington Street*											1
	HUNTINGDON	HUNTINGDON	96.3										
		High Street*	16.2	1970	B1		} 350						
		St. Mary's Square*		1970	B1								
	PETERBOROUGH	PETERBOROUGH	106.0										
		Bridge Street*	68.8	1976	B1		226						6,7
		Cumbergate (part)*											1
		District Centre*											

1	2	3	4	5	6	7	8	9	10	11	12	13	14
CHESHIRE	CHESTER	CHESTER	865.0										
		Bridge Street (part)	115.4	1981	B1	D	130	2	270	760	2	d,h	4,6,13
		Bridge Street (part)	60.5	1981	B1	D	50	1			1	e,h	
		Eastgate Street (part)		1981	B1	D	106	2			4	d,h	
		Eastgate Street (part)		1981	B1	D	90	1			2	e,h	
		Fargate Street		1981	B1		216	1				f	
		Frodsham Street		1981	B1	D	236	1				f	
		Grosvenor Precinct		1966	A2		658						
		Market Square		1981	B1	D	40	2				d,h	3
		Mercia Square		1973	A2/B1		98						
		Music Hall Passage		Pre1600	B1	A,S,D	50						
		Northgate (part)		1972	C	D	40	2				ea,g	
		Northgate (part)		1981	B1	D	110	1				d,h	3
		St. John Street		1981	B1	D	160	2				f	
		St. Werburgh Street		1981	C	D	200					d,h	
		The Cross		1972	C		30					g	
		The Rows		C13th	A1		1042						
		St. Michael's Arcade		1910	A1		60						
		Weaver Street (part)		1980	B1		22						
		Watergate St.		1981	B1	D	220	2				d,h	
	CREWE & NANTWICH	CREWE	97.0										
		Royal Arcade	50.6	Pre1939	A1	D	34	2	20		2	d,h	4
		Gladstone Street (part)			B1		38				1		
		Rear of Asda			A2		322						
		Victoria Street to Asda			A/B1								4
		Near Co-op	11.2		B		98						
		NANTWICH											
		Barker Street			B	A	174						
		Between Castle & Mill Sts			B	A	115				2		
		Between Churchyard Side & Church Lane			B		172						
		Between The Cullet & Hospital Street			B		40						
		Castle Street			B1	A	70						
		Church Lane			B	A	50						
		Churchyard Side			B1	A	132						
		High Street			B	A	234						
		Hospital Street			B		300						
		Hospital St. to Pall Mall			B1		44						
		Love Lane			B	A	130						

1	2	3	4	5	6	7	8	9	10	11	12	13	14
		(Network Continued)											
		Market Street			B	A	160						4
		Mill Lane			B	A	120						1
		Monks Lane, The Cullet											
		& between			B1	A	402						1
		Pepper Street			B	A	160						6
		Pillory Street			B		198						4
		South side of Civic Cnte			B		20						
		Dog Lane			B1		106						
		West side of Kwiksave	78.4		B		70						
	ELLESMERE PORT	ELLESMERE PORT											
		Marina Walk	60.9		B1		226		0		2		4
		Market Sq. and Street			C		196						1
		Market surround			B1		336						1
		Wellington Road		1979	B1		98						1
	HALTON	RUNCORN	95.6										
		Shopping Precinct	35.7		A2		360				1		4
		WIDNES	56.5										
		Albert Road (part)		1979	B1		116					eb	4
		Gossage Street		1979	B1		44						1
		New Street		1979	A/B1		78				1		1
		Kent Street		1979	B1								1
	MACCLESFIELD	MACCLESFIELD	139.5						80		1		
		Ped. Precinct	42.5		A2		200			270	2		4
	VALE ROYAL	NORTHWICH	106.2						170		1		4
		Apple Market Street	17.9		B1		102						
		High Street			B1		116				1		1
		Leicester Street (part)			B1		22						1
		Market Street			B1		130						
		Timber Lane			B1		122						
		Wilton Street			B1		230						
		Wilton Walk			B1		64						
		WINSFORD	24.9										
		Dingle Walk (part)			A2/B2		144				4		1
		Fountain Court (part)			A2/B2		86						
		Queens Parade (part)			A2/B2		66						4
	WARRINGTON	WARRINGTON	161.8										
		Bridge Street & Horse-	124.6		C		274		0	150	2		4
		market Street (pt)											

1	2	3	4	5	6	7	8	9	10	11	12	13	14
CLEVELAND		(network continued)	567.0										1
		Golden Square	99.2		A2		626						
		Sankey Street (part)	31.2		B1		164						
	HARTLEPOOL	HARTLEPOOL											
		Shopping Centre	147.3		A2								14
	LANGBAURGH	REDCAR											
		High Street	157.0	1973	C	S,A	105		230	120			3,8 5,13
	MIDDLESBROUGH	MIDDLESBROUGH									5	*ec*	3
		Brenthall Street (part)		1978	C		70						
		Cleveland Centre			A2	C,S,A,D	420						
		Corporation Road		1978	C	C,S,A,D	524						
		Dundas Mews			B		182						
		Dundas Street (part)			B		68						
		Dundas Street (part)			A2		116						
		Linthorpe Mews			B		146						
		Linthorpe Road			B1		424						
		Newport Crescent			C		114						
		HEMLINGTON											1
		'1 Unit'			C	C,S,A,D	75						
		Viewley Centre	163.0	1978	A2/B2		140						
	STOCKTON-ON-TEES	STOCKTON											
		High Street		1971	C	S,A	125						1,3,8
		Morton Road											1
CORNWALL	CARADON	POLPERRO	377.0										4
		Village Centre	53.9	1967	B1	A,S	Complete Village	1					
	PENWITH	ST IVES	50.9										4
		Town Centre	9.2	1974	B1	A,S	Complete town centre	1				*i*	
	RESTORMEL	ST AUSTELL	72.3					4				*i*	4
		Aylmer Square	24.3	1965	B2	A,S	48						
		Fore Street		1979	B1		184	1					4
CUMBRIA	ALLERDALE	WORKINGTON	476.0								1		4
			94.0						75				4
			28.1										

105

1	2	3	4	5	6	7	8	9	10	11	12	13	14
	BARROW-IN-FURNESS	St. John's Arcade	75.2	1978	A2		50						4
		E.N.S.W. of above	62.7		B		277						4
	BARROW-IN-FURNESS	The Mall, Duke Street			A		86						
	CARLISLE	CARLISLE	100.8										
		Fisher St. (Sthn part)*	69.9	1978	B1		46	3					
		St. Albans Row*		1978	B1		25	3					
		St. Cuthberts Lane*		1980	B1		80	2					
		Town Hall (area Sth of)*		1977	C		50	3					
	COPELAND	WHITEHAVEN	71.8										1,4
		King Street	26.0	1974	B1		276	1					
	SOUTH LAKELAND	HAWKESHEAD											
		Central Area	91.2	1980	B1		75x65	1					4
		KENDAL											
		Finkle Street	20.9	1978	B1		130	1					4, 1
DERBYSHIRE	CHESTERFIELD	CHESTERFIELD	886.0										
		Elder Way	96.1	1977	C	A,U	110	0	110/150		3	j	4,5
		Knifesmith Gate	68.7	1977	C	S,A	190				1	k, z	1
		Burlington Street		1977	B1		155				2	j	1
		High Street		1977	B1		55				2	j	
		Irongate		1980	B1		70					j	
		Packers Row (part)		1977	B1		50					j	1
		Packers Row (part)		1974	B1		30					j	
	DERBY	DERBY	219.3										
		Eagle Centre	215.3		A2		525		0	600	5		5,13
		East Street			B		200				2		
		Sadlers Gate		1970	B		125				2	m	
		Green Street					120						
		Audley Centre*			A2								
		Duckworth Square*			B								
DEVON	EAST DEVON	EXMOUTH	896.0										
		Magnolta Centre	95.1	1979	A2								4
		Associated Streets	24.1	1979	B								

1	2	3	4	5	6	7	8	9	10	11	12	13	14
	EXETER	EXETER	95.6 / 89.5										
		Goldsmith Street				T	65		150		5		4,5
		Guildhall Centre		1979	A/B		200			150	3		1
		High Street		1975	C		330				4	n	3,6,8
		Princesshay & adjacent		1950's?			220						1
		Waterbeer Street					110						
NORTH DEVON		BARNSTAPLE	69.8 / 16.6										
		High Street											
PLYMOUTH		PLYMOUTH	239.3 / 229.5										
		Frankfort Gate											
		Old Town Street		1977	A								
		Plumton -The Ridgeway		1977	B								
		& associated streets											
SOUTHAMS		TOTNES	59.7 / 5.5										
		Fore Street											
TEIGNBRIDGE		TEIGNMOUTH	91.8 / 11.8										
		Boscawon Place		1980	B1								1
DORSET	BOURNEMOUTH	BOURNEMOUTH	553.0 / 153.4 / 139.7										
		Commercial Road		1973	C		175	1/4	150	1100	1		4; 6,13
		Old Cristchurch Road		1973	C	A,S,U	175	1/4			0		3,8
		The Arcade		1900-20	A1	A,S,U	50	3/5			1		3,8
		Westlane Arcade		1900-20	A1		50	3/5			1		
		Burlington Arcade		1900-20	A1		50	3/5					3
	WAREHAM	WAREHAM	4.2										
		Rempstead Centre		1973	B2		75	4					
	POOLE	POOLE	106.7 / 104.0										
		Arndale Centre		1969	A2		250x2	5				o	
		High Street		1973	B1		500	2/4				ed	6
	WEST DORSET	DORCHESTER	74.0 / 12.9										
		Hardye Arcade		1967	A2		50	4				ee	
		South Street		1976	C		240	2/4					
		Tudor Arcade		1960	A2		50	4/5					
		SHERBOURNE	6.6											
		Sutton Yard		1973	B2		50	5				o	

1	2	3	4	5	6	7	8	9	10	11	12	13	14
	WEYMOUTH & PORTLAND	WEYMOUTH St. Mary Street	54.6 40.0	1970	B1		450	2					
	WIMBOURNE	WIMBOURNE Crown Mead	51.3 4.7	1980	B2		100	4				ef	4
DURHAM	CHESTER-LE-STREET	CHESTER-LE-STREET Front Street (part)	608.0 49.9 20.2	1976	C	C	570	1/4	280	480	4		4,5
	DARLINGTON	DARLINGTON	94.2 84.0										
		Blackwellgate (Nth side)		1977	C	C,A	85	1/4				p	3,8
		Blackwellgate (Sth side)		1977	C	C	75	1/4				q	3,8
		Bondgate (part)		1977	C	C,U	35	2/4				r	3,8
		Bondgate (Sth side)		1977	C	C,U,A	210	1/4				p	3,8
		High Row		1977	B1	A	240	1/4				p	
		Horse Market (part)		1977	C	A	150	1/4				p	
		Northgate (part)		1977	C		120	2/4				q	
		Northgate (part)		1973	C		110	1/4				t	
		Post House Wynd		1973	B1		120	1/4				q	
		Prebend Row (West side)		1977	C	C,U	60	1/4				p	3,8
		Skinnergate		1977	C	A	80	1/4				p	1
		Tubwell Row (part)		1977	C	C,U,A		1/4					1,3,8
		Crown Street (part)		1977	C	C,U	60	2/4				r	3,8
	DERWENTSIDE	Court Arcade*		Pre1939	A1			3				s	4
		East Row*		1977	C			1				p	1
		Prospect Place*		1977	C	A	50	2				p	1
		Queen Street*		1968	A2			5				s	
		West Row (West side)*		1977	C	C,U,A		1/4				q	
		STANLEY	91.5 41.8						20		5		
		Albert Street		1973	B1		18					x	1
		Anthony Street		1973	B1		10					x	1
		Clifford Road		1973	B1		250	4				x	1
		Front Street		1973	B1		450	4				x	1
		Mary Street		1973	B1		2					x	1
		Ritson Street		1973	B1		25					x	1
		Scott Street		1973	B1		18					x	
		Thornleyholme Terrace		1973	B1		8					x	
		Co-operative Street*		1973	B1		8					x	1
		Laxey Street*		1973	B1		18					x	1

1	2	3	4	5	6	7	8	9	10	11	12	13	14
DURHAM	DURHAM	DURHAM											
		Elvet Bridge	81.6	1975	B1	S	195	2/4	0	125	2		4
		Fleshergate	21.2		B1	S	55	?			3		5,13
		Framwellgate			B1		70	2					
		Framwellgate Bridge		1975	C	A	115	2			2		
		Millburngate			A2		100						
		Millburngate Centre			C	A,U	225	2/4					3,8
		North Road			B1	A	120						
		Saddler Street		1975	B1		145					v	
		Silver Street											
	EASINGTON	PETERLEE	108.1										
		Yoden Way Centre	21.8		A2		558	4	85		3	w	4
		SEAHAM	23.3										4
		Caroline Street (Nth)		1973	B1		64	1					
		Caroline Street (Sth)		1973	B1		73	4	0		5		
		Church Street		1973	B1		178	4					
		Green Street		1973	B1		28	4/5					
	SEDGEFIELD	NEWTON AYCLIFFE	92.1										4
		Beveridgeway Centre	20.1		B2		294	4	170		1		4
		SPENNYMOOR	19.0										
		Cheapside		1978	C		200	1			1		4
		High Street		1978	C		200	1			1		4
	WEAR VALLEY	CROOK	65.3										
		Hope Street	21.4	1973	B1	S	250	1/4			1		4
EAST SUSSEX	BRIGHTON	BRIGHTON	650.0										
		Churchill Square	166.1	1975	B2		1200	5	550	480	7		4
		Imperial Arcade	153.5	1935	A1		60	3			4		4,5
		Western Road		1975	C	T,S,A	600	1/4			3		3,8
		Dukes Lanes		1980	B2		80	4					
		The Lanes			B2		450	3/4					
		Ditchling Road			B2		250	5					
		Gardner Street		1974	B1		140	1/4				y	
		Kensington Gardens		C.19th	B1		134	3/4					
		Regent Arcade		1960	A2		74	3			1	x	
	EASTBOURNE	EASTBOURNE	70.1						134	67	1		1,4,5
			61.8										

109

1	2	3	4	5	6	7	8	9	10	11	12	13	14
	LEWES	Arndale Centre		1980	A2		140	5					1
		Terminus Road (part)		1973	B		160	2/5					1
		Terminus Road (part)		1978	C			4					
		Langney Centre	71.7	1973	A2		135	5					4
		NEWHAVEN											
		High Street	9.8	1980	B		140	4			2	*eg*	1
	WEALDEN	HAILSHAM	108.0								1		
		Vicarage Field	51.5	1965-66	B		50	4	440				4
ESSEX	BASILDON	BASILDON	1354.0										4
		East Square	129.2	1963	B2		35	4			2		3,4
		East Walk	77.1	1963	B2		180	4			2		
		South Gunnels		1963	B2		100	4					
		Town Square		1963	B2		250	4					
		CORRINGHAM											
		Town Centre		1965	B2		65	4					4
	BRAINTREE	BRAINTREE	93.2										
		Market Street	24.4	1978	B2		60	4					4
		New Street		1978	B2		75	4					1
		WITHAM											1
		Newlands Centre	16.9	1969	A2		100	4					4
	CHELMSFORD	CHELMSFORD	122.6										
		Conduit Street	56.7	1977	B1		30	5					4
		Cornhill			A2		100	4					4
		Exchange Way			A2		100	4					1
		High Chelmer			A2		100	4					
	COLCHESTER	COLCHESTER	117.7										
		Culver Street East	71.2	1974	B1		70	5					4
		Culver Walk		1974	B2		160	5				*z*	
		Eld Lane		1978	B1		210	5				*aa*	
		Lion Walk		1974	B1		95	5				*aa*	
		Long Wyre Street		1978	B1		130	4/5				*aa*	
		Short Wyre Street		1978	B1		55	2					
	HARLOW	HARLOW (Old Town)	77.7										
		High Street		1972	B1		150	4					4
		Staple Tye Centre	77.4	1960	A2		8000m²	5				*ab*	

1	2	3	4	5	6	7	8	9	10	11	12	13	14
		Stow-Bush Fair		1956	B2		10000m²	4					4
		HARLOW (New Town)											
		The Stow	40.3	1962	A2		100	4				ac	4
		Town Centre	13.5	1958	A2/B1		70000m²	4				ac	4
	MALDON	MALDON											
		King Edwards Walk	162.3	c.1970	A2		75	4					4
		SOUTH WOODHAM FERRERS												
		Town Square	159.3	1978	A2			4					4
	SOUTHEND-ON-SEA	SOUTHEND-ON-SEA											
		High Street		1978	B1		300	4					1
		Sutton Street	124.6										
	THURROCK	CHADWELL ST MARY											
		Defoe Parade		1974	B1		50	4					4
		GRAYS												
		High Street		1973	B1			4					1
		West Mall			A2			4					4
		SOUTH OCKENDON												1
		Derwent Parade		1979?	B1		100	4					4
GLOUCESTERSHIRE	CHELTENHAM	CHELTENHAM	463.0								0 0		5,13
		High Street (part)	79.8 / 71.0	1977	B1		120	2/4	340	1780	1 6		
	GLOUCESTER	GLOUCESTER	90.1						30	156	0 1		5
		Bell Walk	86.9		A2		89				1		
		The Forum			A2		48						
		Market Way			A2		60						
		Blackfriars			B		102						
		Clarence Walk			A2		40						
		Constitution Walk (part)			B	S	74						
		Constitution Walk			B	S	26						
		Greyfriars (part)			B		130						
		Greyfriars (part)			B		80						
		Kings Square			B		186						
		Kings Walk (part)			A2		72						
		Kings Walk (part)			B		114						
		Queen Street			A2	S	52						
		Berkeley Street			B		112						14

1	2	3	4	5	6	7	8	9	10	11	12	13	14
GREATER LONDON			7344.8 158.7										
	BARKING	BARKING											
		East Street	305.6	1980	B1		120	1/4/5		164	1 0 0 3	*aa*	5 1 5
	BARNET	BARNET											
		Brent Cross		1976	A2		700	4					14
		COLINDALE		1974	B2		135	4					
		Grahame Park											4
		HENDON											1
		Sentinal Square	217.9	1972	B2		60	4					1
	BEXLEY	ERITH											
		Cross Street*											
		Pier Road*		1970?	B2								
		Shopping Centre*		1970	B2								
		THAMESMEAD											
		N. Shopping area	274.8					5					5
	BRENT	KILBURN											
		Kilburn Square		1967-68	B2		40	5					
		NEASDEN		1975	B1		67	4					
		Town Centre											
		S. KILBURN		1969/70	B2		60	4					
		Peel Precinct											
		WEMBLEY		1965	B2		43	4			1		4,5
		Wembley Square	306.3										
	BROMLEY	BROMLEY											
		Clarina Road		1979	A2		125	4					1
		The Mall Precinct*		1970	A2		350	4					
		ORPINGTON		1973	A2			5					
		The Walnuts											
		ST MARY CRAY		1979	B1			1					1
		High Street	196.7										4,5
	CAMDEN	CAMDEN TOWN											
		Brunswick Centre		1972	B2		100	5					
		Cosmo Place		Pre1900	B2		70	3				*ae*	
		Invermess Street		1974	B1			2					
		FITZROVIA											
		Fitzroy Square											
		HAMPSTEAD		1975	B1			1					
		Oriel Place											

1	2	3	4	5	6	7	8	9	10	11	12	13	14
		Perrins Court											
		HOLBORN											
		Lambs Conduit Passage		Pre1900	B2			3					11
		Lambs Conduit Street		1977	B2		45	1					11
		Little Turnstile		Pre1900	B2			3					4,5
		Sicilian Avenue		1965	B2		60	3					11
	CITY	Woburn Walk	4.3	Pre1900	B2		60	3					11
		The Arcade		Pre1939	A1								11
		Barbican		1959	B2								11
		Bow Lane											
		Cheapside											
		Paternoster Square		1961	B2						1		11
		Tower Precinct	334.0	1959	B2					1	2	1	4
	CROYDON	CROYDON											
		Parade by Leon House		Post1945	B1		180?	3					
		St. Georges Walk		Post1945	A2		100?	4					
		Whitgift Centre	229.3	1970?	A2		300?	5				*af*	5
	EALING	Acton Precinct	226.0	Mid1960s	B2		75	5	0				5
	ENFIELD	EDMONTON											
		Edmonton Green Shopping Centre		1970-73	A2/B2			5			1		5
	HAMMERSMITH & FULHAM	Kings Mall	181.9	1978-79	A2		230	5		215	2		4,5
		Shepherds Bush Centre	235.6	1973-74	A2/B2		170	4					
	HARINGEY	WOOD GREEN											
		Dovecot Avenue		1969	B1		30	4					
		Shopping Centre	203.4		A2								
	HARROW	HARROW											
		Station Road		1979-80	B/C	S,U	100?	0	150		1		4,5
		PINNER											
		Bishops Walk		1978	B2		135	4					
		SOUTH HARROW											
		Sherwood Road		1979	B1		45	4					

1	2	3	4	5	6	7	8	9	10	11	12	13	14
	HAVERING	ROMFORD											
		North Street Precinct	247.0								3		5
		Quadrant Arcade		1969-72	B1		190	4			2		1
				Pre1939	A1		125	3/4					
		South Street	236.4	1978	B1	S,U	130	4/5					1,4
		The Liberty		1971	A2/B2		780	5					
	HILLINGDON	UXBRIDGE											
		George Street*	206.7										1
		High Street*											1
	HOUNSLOW	FELTHAM											
		The Centre	182.2	1965	B2		170	4			1		4,5
	KENSINGTON & CHELSEA	Brompton Arcade	139.0	Pre1900	A1		52.2	3			2		4,5
	KINGSTON-UPON-THAMES	Alderman Judge Mall		Post1970	A2		61	5					
		Church Street (part)		Early70s	B1		68	2					
		Crown Arcade		InterWar	A1		30	4/5				ag	
		Crown Passage		1970	B1		102	2/4/5				ah	
		Eden Walk		C19th			145	5				ag	
		Market Place		Early70's	B1		123	2					
		Alliance Development	300.3	Post1970	A2		37	5					4,5
	LAMBETH	BRIXTON											
		Turnstall Road	262.9	1980	B1		30	1					1
	LEWISHAM	CATFORD											
		Catford Centre		1973	A2/B2		30	4/5					4,5
		Milford Road			A2/B2		150	5					
		Winslade Way*		1973	A2/B2			3					1
		DEPTFORD											
		Kingfisher Square		1972	B2		55	4					
		High Street		1975	C		400	2					
		Tanners Hill		1975	B1		60	2					
		Woodpecker Centre		1975	B2		80	4					
		LEWISHAM											
		Town Centre	176.8	1977	A2		154	5				ai	
	MERTON	Wimbledon Arcade		c.1900	A1		75	3					4,5

1	2	3	4	5	6	7	8	9	10	11	12	13	14
	NEWHAM		231.6										
		EAST HAM											
		Shopping Hall		Pre1945	A1		200	3/4	52	80	3		5
		Queens Road		1970	A2		150	4			1		
		Shopping Centre		1972	A2	S	350	4/5					5
	REDBRIDGE		237.6										
		ILFORD											
		Centre Way			A2		45	1/5	75	520	1		5
	RICHMOND-UPON-THAMES		172.7										
		RICHMOND											
		Golden Court		1975	B1		33	3					
		Lower George Street		1974	B1		40	3/4/5					4
	SOUTHWARK		253.0										
		NEWINGTON											
		Elephant & Castle		1960?	A2/B2?		180?	5					1,11
		PECKHAM											
		Holley Grove					203?						
		Rye Lane			C		833?				1		3
	SUTTON		169.2										
		CARSHALTON											
		High Street		1978	B1								1
		SUTTON											
		High Street	158.8	1979	B1								5
	TOWER HAMLETS												
		Stroudley Walk		1979	B1		72	4			1		
		Vernon Road (part)		1979	B1	U	32	1			1		4
	WALTHAM FOREST		233.2										
		High Street	296.4	1976-80	B1		320	5	164	175	2	*eh, a[i]*	4,5
	WANDSWORTH												
		BALHAM											
		Hildreth Street		1978	B1	C	20	2/4			1		
		CLAPHAM JUNCTION											
		St. Johns Road		1977	C	S	310	1/4		150		*ak, a[d]*	3
		WANDSWORTH											
		Arndale Centre		1970-74	A2		87	3/4					
	WESTMINSTER		231.4										
		Adelaide Street		1979	B1		70	5					4,5
		The Arches		C19th	A1		81	3					
		Craven Passage					75						
		Villiers Street		1970	B1		200	1				*am, an*	11
		Watergate Walk		C19th	B2		70	3				*aq*	11

1	2	3	4	5	6	7	8	9	10	11	12	13	14
		Barrett Street		1980	B1		38	3				*ao*	
		Gees Court (part)		1980	B1	)	62	3				*ao*	
		Gees Court(part)		Pre1939	B2	)						*as*	
		Jason Court											
		Oxford Street		1972-74	C	T,S	37	5				*at*	9,11
		St. Christophers Place		C19th	B2		1665	3					
							88	3					
		Bateman Buildings		C19th	B2		75	3					
		Bishops Court		C19th	B2		45	3					
		Chichester Rents		C19th	B2		35	3					
		Star Yard		C19th	B2		50	3					
		Broad Court		C19th	B2		100	3				*an*	11
		Burlington Arcade		C19th	A1		70	3					
		Carnaby Street		1972	B1		216	2				*ap*	11
		Fouberts Place (part)		1972	B1		50	3/4				*ap*	11
		Fouberts Place (part)		C19th	B2			3					
		Ganton Street (part)		1972	B1		45	3/4				*ap*	11
		Ganton Street (part)		C19th	B2			3				*ar*	
		Cecil Court		C19th	B2		79	3					
		Chapel Place		C19th	B2		60	3					
		Charlotte Place		C18th	B2		43	3					
		Church Place			B2		43	3					
		Covent Garden		1979-80	B2		115	3/4					
		Cranbourne Alley		C19th	B2		20	3					
		Crown Passage					75						
		Devereux Court		C19th	B2		75	3					
		Eagle Place		C19th	B2		45	3					
		Elizabeth Court		C19th	B2		40	3					
		St. Ann's Lane		C19th	B2		65	3					
		George Court		C19th	B2		40	3					
		Goodwins Court		C19th	B2		45	3					
		Hop Gardens		C19th	B2		60	3					
		Greens Court		C19th	B2		35	3					
		Hanover Place		C19th	B2		35	3					
		Heddon Street		C19th	B2		45	1					
		Jones Street		C18th	B2		30	3					
		Kemp's Court		C20th	B2		18	3					

1	2	3	4	5	6	7	8	9	10	11	12	13	14
		Knightsbridge Green		C19th	B2		80	1					
		Lancashire Court		C19th	B2		50	3					
		Langley Court		C19th	B2		30	3					
		Landsdowne Row					69						
		Leicester Court		C19th	B2		50	3					
		Leicester Square		1975–80	B1	A	180	2					
		Little Marlborough St.		1972	B1		65	3/4				ap, am	11
		Marlborough Street		C19th	B2		125	3				am	
		May's Buildings		C20th	B1			3					
		Meard Street		C18th	B2		40	3					
		Newport Court		C19th	B2		50	3					
		Palmer St./Brewers Grn.		1977	B2		110	5					
		Park Close		C20th	B2		50	3					
		Park Mansions Arcade					43						
		Piccadilly Arcade											
		Piccadilly Place		C19th	B2		25	3					
		Pickering Place		C18th	B2		25	3					
		Quadrant Arcade					44						
		Rolls Passage		C19th	B2		65	3					
		Royal Arcade					33						
		Royal Opera Arcade		C19th	A1		75	3				am	
		St. Albans Street		C19th	B2		50	3					
		St. Anne's Court		C19th	B2		90	3					
		St. Martin's Court					80						
		Seaforth Place		C19th	B2		60	3					
		Sedley Place		C19th	B2		90	3					
		Shavers Place		C19th	B2		70	3					
		Shepherds Place		C19th	B2		55	3					
		Silver Place		C19th	B2		35	3					
		South Molton Street		1977	B1		218	3/4					11

1	2	3	4	5	6	7	8	9	10	11	12	13	14
		Tennison Court		C18th	B2		30	3					
		Tisbury Court		C19th	B2		30	3					
		Vandon Passage		C19th	B2		50	3					
		Victoria Arcade					25						
		Walkers Court		C19th	B2		35	3				*an*	
		Westminster Cathl. Piazza		1977	B2		77	4					
		Angle Court*		C19th	B2		63	3				*an*	
		Grand Arcade*						3/4					
		Lownds Court*		C19th	B2			3/4					
		Marlborough Court*		C19th	B2								
GREATER MANCHESTER			2727.0										
	BOLTON		259.0										
		BOLTON	152.3										
		Arndale Centre		1971	A2		80	5	160		1		1,4,5
		Newport Street		1970	B1		160	4				*au*	5
		Oxford Street		1970	B1		80	4				*au*	4
		Victoria Square		1970	B1		290	4			1	*au*	4
	BURY		174.0										
		BURY	66.8										
		Haymarket Precinct		1969	B2		380	4	50		3	*au*	5
		PRESTWICH											
		Longfield Centre	30.6	1971	A2			4		122	2		4
		WHITEFIELD											
		Elms Square	21.8	1968	A2			4	146	210	1		4
	MANCHESTER		542.0										
		CHEETHAM HILL											
		Cheetham Parade Precinct		1970	B2		175	4			1		5
		CHORLTON											
		Chorlton Place Precinct		1971–72	B2		160	4			1		4,5
		MANCHESTER											
		Arndale Centre	526.3	1976–77	A2		750	5		0/20/150	2		4,5,13
		Brown Street		1971	B1		60	5			2	*aaa*	
		Market Centre		1967	A2		160	5			2	*aaa*	
		King Street (part)		1976	B1		170	2					
		Market Place Precinct		1972	B2		325	5					
		Parsonage Lane		1973–74	B1		25	2					
		Piccadilly Plaza		1962	A2		325	4/5					
		Royal Exchange Arcade		Pre1939	A1		55	4					
		St Anns/Barton Arcades		Pre1939	A1		20/55	4					1

1	2	3	4	5	6	7	8	9	10	11	12	13	14
		MOSS SIDE											
		Moss Side Centre		1973-74	A2			5					4,5
		SALE											4,5
		School Road (part)			B1		250				5		
		Town Square Precinct		1965	A2		220	4					4,5
		WYTHENSHAWE	224.0										4,5
		Wythenshawe Centre	23.2		A2								4
	OLDHAM	FAILSWORTH											
		Failsworth Precinct		1975	A2			4			1		4
		OLDHAM	103.7										
		Hilton Arcade		1891	A1		60	3/4	100/140		1		
		Market Avenue		1891-1911	A1		60	3/4			1		
		St. Peters Precinct		1968	A2		245	5			1		4
		ROYTON	20.3										
		Market Street Precinct		1973	A2		85	4					
	ROCHDALE	MIDDLETON	203.0										
		Arndale Centre	52.9		A2		285		5				1,4
		Market Street*											4
		Saddler Street*											
		Taylor Street*											1
		ROCHDALE	90.3										
		Reed Hill					30		25		3		
		The Baum					30				1		1
		The Butts Arcade		Pre1939	A1		125				1		
		Yorkshire Street Centre			A2		310						5
	SALFORD	ECCLES	279.0										
		Church Street	37.8	1971	B1		234	4			5		4
		Church Street Precinct		1966	A2		460	5	50		3		
		IRLAM	20.4										
		Liverpool Road Precinct		1968	A2			4					4
		SALFORD	127.6										
		Ordsall District Centre		1977	B2			4					
		Regent Square(The Piazza)*		1963	B2			4					1
		Salford Precinct*		1970	A2			5					
		SWINTON	39.6										
		Swinton Parade Precinct		1966	A2			5					4
		WALKDEN											
		Walkden Precinct		1968-74	A2			4					4

1	2	3	4	5	6	7	8	9	10	11	12	13	14
	STOCKPORT	BRAMHALL	292.0										5
		Bramhall Precinct			B2			4					4
		CHEADLE HULME	59.8										4
		The Precinct			B2			4					4
		MARPLE	23.5										4
		Market Street		1978	B1			4					4
		STOCKPORT	137.2										
		Merseyway Precinct			A2/B2		275	4	60		14	2	1
		Prince's Street			B1		235	4			2	1	5
	TAMESIDE	ASHTON-UNDER-LYNE	221.0										4
		Warrington Street Prect.	48.2	1967	A2		392	4	20		3	1	1
		DENTON	38.0										
		Market Street		1969	B1			4					4
		Victoria Street		1969	B1			4					
		DROYLESDEN	24.1										1
		Queens Precinct		1972	A2			4					4
		DUKINFIELD	17.2										
		Town Lane		1972	B2			1/4					
		HYDE	36.8								1		1
		Market Place Precinct		1966	A2		310	4	130		1	2	4
		STALYBRIDGE	22.7										4
		Bennett Street		1978	B1			4					1
		Canal Street		1978	B1			4					1
		Grosvenor Street		1978	B1			4					1
		Kenworthy Street		1978	B1			4					1
		Melbourne Street		1978	B1			4					
	TRAFFORD	ALTRINCHAM	227.0										4
		George Street	40.1		B1		246				3		
		Grafton Mall			A2		232	4	15		1		
		STRETFORD	53.8										4
		Arndale Centre			A2								
	WIGAN	LEIGH	306.0										5
		Bradshawgate Arcade	45.8	1975	A2		60	4					
		WIGAN	80.4								1		4
		Makinson Arcade		1898	A1		95	3	110				5
		Wigan Centre Arcade		1974	A2		140	5					4
		Royal Arcade		1927	A1		60	3					5
HAMPSHIRE	BASINGSTOKE		1370.0										
			103.2										

120

1	2	3	4	5	6	7	8	9	10	11	12	13	14
	BASINGSTOKE	London Street	52.1	1976	B1						2		3,4,5
		New Market Square		1966	B2		175	4		315	2		
		New Town Centre		1968–70	A2/B2		42x42; 800	4; 5			1	a)	
	FAREHAM	West Street	80.3	1977	B1		210	2/4					4,5
		Westbury Precinct	78.3	1975	A2		1102	5					4,5
	HAVANT & WATERLOOVILLE	Greywell*	109.0	Pre1970	B2		150	4					
		Market Parade*	107.8	Pre1970	B2		50	4					4,5
		Park Parade*		1972–78	B1		150	4					
		Wellington Way*		1965	B2		150	4					
	NEW FOREST	HYTHE High Street	131.3	1980	B1		130	4			1		4
		LYMINGTON Quay Hill	34.3	Pre1939	B1			1			1		
		Quay Street		Pre1939	B1	}	100	1					
	PORTSMOUTH	Arundel Place	197.0								2		1,4,6
		Arundel Street	184.7							40	1		
		Commercial Road		1972	C		105				2		1,3,6,13
		The Tricorn									1		1
		Crasswell Street											1
		Swan Street			A2								1
		Russell Street*											1
		Sussex Street*											4
	SOUTHSEA	Palmerston Road		1976	C		365						1
		Stanley Street		1979	B1								1
	RUSHMOOR	ALDERSHOT Union Street	74.5	1978	B1		40	1			5		5
		Wellington Street	30.6	1978	B1		35	1	70	70	1		1,4
		Wellington Centre											
	FARNBOROUGH	Kingsmead*	39.6	1975	B1								4
		Queensmead*			B1								
	SOUTHAMPTON	SHIRLEY	214.8										5

1	2	3	4	5	6	7	8	9	10	11	12	13	14
		Cannon Street	206.6	1976	B1		27	4	150	670	1	1	1
		SOUTHAMPTON											
		Above Bar Precinct									2	0	4,13
		East Street		1971	B1		215	4					1
		Redcar Street		1977	A2		160	4					1
		TOTTON											
		Library Road		1972	B1/B2		85	4					4
	TEST VALLEY	Manchester Street	79.5	1961	B1		55	3					
		ANDOVER	25.3										
		Chantry Way		1971	B2		240	4/5			4	3	4,5
		Union Street		1967	A2/B1		120	4			2	2	
		ROMSEY	9.8										
		Dukes Mill		1976	B2		90	4			3		
	WINCHESTER	WINCHESTER	89.7										
		High Street	28.1	1973	B1		200	3/4	0	550	3	2	4,5,13
		Kings Walk		1975	A2		60	4					
HEREFORD AND WORCESTER	BROMSGROVE	BROMSGROVE	562.0										
		Mill Lane	77.1										1
	HEREFORD	HEREFORD	46.5										
		Church Street	45.1				100				1		
		High Town			B		280		400	620	1	1	4,13
		Eign Gate					100				1		
		St. Owens Gate*	33.2										1
	LEOMINSTER	LEOMINSTER											
		Drapers Lane	9.2										
	REDDITCH	REDDITCH	40.8										
		Evesham Street*	37.6		C?								4,6
		Shopping Centre*		1976	A2								1
	WORCESTER	WORCESTER	73.6										
		The Shambles	70.9	1978	C		155		200	750	2	2	4,6,13
	WYCHAVON	DROITWICH	82.1										4
		Miller Street*	12.2										1

1	2	3	4	5	6	7	8	9	10	11	12	13	14
	WYRE FOREST	St. Andrews Road*	84.9										1
		Westcroft Street*	46.7										1
		KIDDERMINSTER	922.0										
		Bull Ring	71.0	1972									1
HERTFORDSHIRE	BROXBOURNE	BROXBOURNE											
		The Precinct	26.0	Late60's	B2		80	4			1		5
		HODDESDON											
		Falkdon Walk			B2			4			1	*aa*	4
		Tower Centre	14.3	Late 60's	B2		130				0		
		WALTHAM CROSS											
		The Pavilion		Late 70's	A2			4/5			1		4
	EAST HERTFORD-SHIRE	BISHOPS STORTFORD	102.1										
		Jackson Square	21.0	Late 70's	A2								4
		HERTFORD	19.5										
		Maidenhead Street (part)			B1	A						*ba*	4
		The Arcade		Pre1939	A1		60	3					4
	NORTH HERTFORD-SHIRE	SAWBRIDGEWORTH	7.0										
		Town Centre											
		HITCHIN	99.3										
		Arcade*	27.8	1922-23	A1		70	4					5
		Churchgate*		Late 70's	A2		75	4					4
		LETCHWORTH	30.1										
		The Arcade*		1922-23	A1		60	3					4
		Commerce Way*		Late 70's	A2		290	4					
		ROYSTON	8.2										
		Angel Pavement		1967	B2		55	4					4
	ST ALBANS	ST ALBANS	121.4										
		French Row	50.7	1975	B2		50	4					5
		Heritage Close		1975-76	B1		65	4					4
	STEVENAGE	STEVENAGE	67.0										
		The Forum	66.7	Early50's	B2		120	4	0	246	6 2		5
		Market Place		Early50's	B2		110	4			2		4
		Park Place		Early50's	B2		100	4					
		Queensway		Early50's	B2		330	4					
		Town Square		Early50's	B2		70	4					

123

1	2	3	4	5	6	7	8	9	10	11	12	13	14
	THREE RIVERS	OXHEY											
		St. Andrews Road	80.2	Mid 60's	B1		120	4					5 4
	WATFORD	WATFORD	78.1 76.7							545/470	5		4,5,6
		Charter Place		1976	A2/B2		325	5					
		Clarendon Road		1972	C		110	1					
		High Street (part)		1962-	} B1/C		275	4			1		
		High Street (part)		1980			710	1			1		
		Market Street		1966	C		130	1/4					
		Queens Road		1979	B1/C		200	4			0		
		Vicarage Road		1968	B1		70	4					
		Water Lane		1980	C		60	1					
		Hempstead Road		1972	B1		130	4					
	WELWYN HATFIELD	HATFIELD (Old Town)	} 93.7								3		5
		Salisbury Square		1973	B2		120	4		90			
		HATFIELD (New Town)	} 25.0								2		4
		Town Centre		1967-68	B2		1095	4			-		
		The Arcade									1		
		Market Place									1		
HUMBERSIDE	BEVERLEY	BEVERLEY	838.0										
		Toll Gavel	101.3 16.7		B								
	GRIMSBY	GRIMSBY	95.7 93.0										
		Bull Ring*											
		Riverhead Centre*			A2								4 1
	KINGSTON-UPON-HULL	HULL	285.5 280.0										
		Prospect Centre			A2		125	5	300	200	0		4,13
		Whitefriargate			B		200				0		
	NORTH WOLDS	BRIDLINGTON	65.4 25.8										
		King Street			B1						0	*bb*	4
	SCUNTHORPE	SCUNTHORPE	70.9 69.7									3	4
		High Street (part)*			A								1
		High Street (part)*			B								1
		John Street*											1
		Manley Street*											1

1	2	3	4	5	6	7	8	9	10	11	12	13	14
ISLE OF WIGHT	MEDINA	Market Hill*											1
		Southgate*											1
		COWES											
		The Cut	109.0	1975	B1		43	0					
		High Street (part)	64.4	1975	B1	S	90	0					
		Shooters Hill	18.5	1975	B1	S	70	2/3					4
		RYDE	21.9										
		Anglesea Street		1976	B1	S,D	110	0					4
		High Street (part)			B1	S,D	180	0					
		High Street (part)			C	S,D,A		0					
KENT	ASHFORD	ASHFORD	1396.0										
		Upper High Street	78.9	1978	B1		95	4	500	590	3		1,5
		Tufton Centre	34.8	1975	A2		207	4			1		
	CANTERBURY	CANTERBURY	109.9										
		Burgate	30.1	1973	B1	S	110	0	20	625/700	2		5,8,13
		St. Georges Street		1973	B1		110	4			1		1
	DOVER	DEAL	99.0										
		High Street (north)	24.4	1977	B	A,D,S	105	1	70		2		
		High Street (south)		1970	B		85	2			2		5
	GILLINGHAM	GILLINGHAM	86.7										
		High Street	85.1										1
	MAIDSTONE	MAIDSTONE	120.7										
		Gabriels Hill*	67.7	1974	C	D,S	181	1			1		5
		Pudding Lane*		1978	C		39	0			0		3
	MEDWAY	CHATHAM	138.4										
		Pentagon Shopping Centre			A2		400	5			0	*bc*	
		High Street (part)*	81.8		B1		175	2			0	*bd*	13
	SHEPWAY	FOLKESTONE											
		Sandgate Road (part)	41.8		B2				150		1	*be*	5,6
	SWALE	FAVERSHAM	100.7					100	4			1		1
		Partridge Lane*	14.5	1974?										1

1	2	3	4	5	6	7	8	9	10	11	12	13	14
LANCASHIRE		West Street*		1974?									1
		SITTINGBOURNE	30.6										1
		Crescent Street											
	BLACKBURN	BLACKBURN	1341.0										
		King William Street (pt)	140.5	1977	A2		60	5					5
		Lord Street	99.6	1977	A2		100	5					
		Northgate Jct.		1971	A	A		1					
		DARWEN	28.7										
		School Street		1972	B1		50	4					1
	BLACKPOOL	BLACKPOOL	151.3										
		Adelaide Street (part)	140.6	Post '74	B1/C	A,T	76	2	400/800	500	2		1,5,13
		Bank Hey Street (part)		Pre '74	B1/C	A,T	144	2					3
		Church Street (part)		Post '74	B		78	2			1		3
		Corporation Street (pt)		1978?	B1		64	2					
		Hounds Hill		1980	A		73	5					
		Hounds Hill Centre		1980	A2		420	5					
		Victoria Street (part)		1978?	B1		195	2					
		Winifred Street (part)		Post '74	B		35	2					
	BURNLEY	BURNLEY	94.8						130		1		1,5
		Brick Street	75.0	1979	B1		50	5			1		
		Chancery Walk		1967-72	B2		45	5				*bf*	
		Fleet Walk		1967-72	B2		40	5				*bf*	
		Howe Walk		1967-72	B2		40	5				*bf*	
		The Mall		1967-72	B2		105	5				*bf*	
		Market Square		1967-72	B2		80	5				*bf*	
	LANCASTER	LANCASTER	123.5						130		5		5
		Cheapside	43.8	1973-76	B1		65	2/4			0		
		Lancaster Gate (part)			B1		12	2/4/5					
		Market Square		1973-76	B1		46	2/4			3		
		Market Street		1973-76	B1		232	2/4					
		Music Room Square		1973-76	B1		58	2/4			5		
		Penny Street (part)		1976	B1		38	2/4					
		St. Nicholas Shopping Centre			B1/B2								
		Sir Simons Arcade					174	2/4/5					
		Sun Street					73						
	MORECAMBE	MORECAMBE	40.1										
		Anderton Street*					23						1
		Euston Road*		1973	C								1 / 3,8

1	2	3	4	5	6	7	8	9	10	11	12	13	14
	PRESTON	PRESTON											
		Market Street (part)	135.1	1975	B1	S	110	1	0	370	3		5
		New Market	95.9	1975	B2	S	110	5					
		Orchard Street		1975	B1	S,T	95	1					8
		Cheapside		1975	B1		100	0					
		Fishergate (part)		1975	C		100	5					
		Harris Street		1973	B1		80	5				bh	
		Lancaster Road (part)		1973	C		70						
		Stoneygate			B2		110	3					
		Tithebarn Street (part)		1969	B1		50	4					5
		St. George's Circus		1965	A2		220	4					
		Birley Street*		1973	B1		80	2			2	bg	5
	ROSSENDALE	RAWTENSTALL	61.8										
		The Centre	21.0	1967-68	A2/B2		155						5
	WEST LANCASHIRE	ORMSKIRK	91.4										
		Aughton Street (part)*	27.0	1980	B		100	4					
		Burscough Street*		1980	C	S	170	4					
		Church Street (part)*		1980	C		150	4					
		Moor Street (part)*		1980	C		900	4					1
		SKELMERSDALE											
		Digmoor*	39.0	1969	A2		170	4					
		Sandy Lane*		1971	A2		106	4					
		Town Centre Concourse*		1973	A2		250	5					5
	WYRE	POULTON-LE-FYLDE	95.0						0				
		Church Street	16.3		B1		70				2		1
		Teanlowe Centre			A2		120						
LEICESTERSHIRE	CHARNWOOD	LOUGHBOROUGH	799.0										
			125.4										
		Church Gate	42.7	1979	B1	S	95	2	250	550		bi	4,5
	LEICESTER	LEICESTER	283.5										
		Clock Tower	276.8	1978	C		45	0			1		1,5,13
		East Gate		1978	C	S	214	2			0		
		Gallowtree Gate		1971	A2	S,C	160	2			0		
		Haymarket Shopping Cntre			C							bj	3,6,8
		Humberstone Gate		1978	C		40	2			3		1

1	2	3	4	5	6	7	8	9	10	11	12	13	14
LINCOLNSHIRE	BOSTON	BOSTON											
		Bargate (part)	503.0		B1		140	3/4/5	300	500			5
		Dolphin Lane	48.8	1980	B1		100	3/5				bk	1
		New Street (part)	25.6	1980	B1		40	3					
	LINCOLN	LINCOLN											
		Cornhill	74.2	1973	B1	S	135	2	20	6n	2		5,13
		Guildhall	71.6		B1		100	4				bZ	
		High Street		1973	B1	S	270	2				bZ	
		Saltergate			B1		125	4					
		Sincil Street			B1		170						
		Watergate			B2		160						
	NORTH KESTEVEN	St. Benedict Square	72.7	1973	B1	S	155	2				bZ	5
		SLEAFORD	7.4										
		Two Streets* }			A								14
		Ped. Precinct*			A2/B2								14
	SOUTH KESTEVEN	STAMFORD	85.3										
		High Street	4.8		B1								
MERSEYSIDE	KNOWSLEY	HUYTON	1659.0										
		Derby Road (Archway Rd. to Huyton Hey Rd.)	198.0	1977	B1		370	4			1		4
		Sherbourne Square	66.1	1965	B2			4			1		4
		KIRKBY											
		Kirkby Precinct	59.7	1961+	B2		90	4	0		5		4
	LIVERPOOL	LIVERPOOL											
		Basnett Street	607.0	1974	B1		120	5			3		4
		Church Alley	590.2	1974	B1		60	2			3		
		Church Street		1974	B1		300	4/5			5		
		Houghton Street		1974	B1		150	2			2		
		Leigh Street (part)		1974	B1		45	1					
		Lord Street (part)		1974	C	A,S	140	2					1
		Richmond Street		1974	B1		80	5					1
		St. Johns Centre & Mkt.		1968-72	A2		380		90/90	80/30	1	bm	3
		Tarleton Street		1974	B1		170	2					
		Whitechapel		1974	C	A,S	200	1					3
		Williamson Square		1970?	B		65	2					1

1	2	3	4	5	6	7	8	9	10	11	12	13	14
		(network continued)											
		Williamson Street		1974	B1		130	2				bo	
		Bold Street		1971	B1		320	2/4					4
		Cases Street		1974	B1		65	2					
		Clayton Square		1974	B		50	3					
		India Building Arcade		Pre1939	A1		70	3				bm	
		NETHERLEY											
		Belle Vale District Ctr.		1973	A2		240	5					4
	ST HELENS		192.0										
		ST HELENS	102.6										
		Barrow Street		1970's	B1		100	1					
		Church Street		1970's	C		180	1			1		
		Hardshaw Street		1970's	B1	A,S	100	1			1		
		Tontine Precinct		1970's	A2		50	5				bp	3,8
	SEFTON		307.0										
		BOOTLE	73.6										
		New Strand Precinct		1968+	A2		500	4	0		3		4
		LISCARD											
		Liscard Precinct		1965	B2		240	4/5		120	2		4
		Liscard Road		1979	B1		75	2					
		SOUTHPORT	80.9										
		Cambridge Arcade		Pre1939	A1		90	4		85			4
		Wayfarers Arcade		Pre1939	A1		95	4				bq	
	WIRRAL		355.0										
		BIRKENHEAD	134.8										
		Grange Precinct		1976	A2		170	4/5	100		10		4
		Grange Rd. & Assoc. Sts.		1978	B1		265	2			1		
NORFOLK			624.0										
	BRECKLAND		76.4										
		DEREHAM	9.2										
		Nelson Place		1971	B2		65	4				br	4
		THETFORD	13.6										
		King Street*		1973	B1		280	4					4,14
		Riverside Walk*		1973	B2		150	4					1
		Tanner Street*		1973	B1		75	4					1, 14
	BROADLAND		89.8										
		WROXHAM											
		The Broads Centre		1972	B2		70	4					4
	GREAT YARMOUTH		71.4										
		GREAT YARMOUTH	47.6										
		Central Arcade		Pre1939	A1		130	4					4
		Market Gates Precinct		1975	A2		275	5					

1	2	3	4	5	6	7	8	9	10	11	12	13	14
	NORTH NORFOLK												
		WELLS	74.1										4
		Staithe Street	2.3	1977	B		90	2				bs	4
	NORWICH	NORWICH	121.7										4,5,7
			117.6										3,6,8
		Gentlemen's Walk		1975	C	T,D	250	2	325	675	3		
		Hay Hill		1971	B1		120	3/4			2	bu	
		Royal Arcade		Pre1939	A1		75	2			3		
		White Lion Street		1971	B	S	100	0			2		
		Lower Goat Lane		1970	B1		90	3/4					1
		Dove Street		1970	B1		75	1					
		London Street		1967	B1		180	3/4				bt	
		Anglia Square*		1968	B2		150	4					4
	SOUTH NORFOLK		80.7										4
		DISS	4.4										
		Mere Street		1976	B		130	2				bt	4
	WEST NORFOLK		110.0										4
		DOWNHAM	3.5										
		Wales Court			B2		90	4	145		4		
		KINGS LYNN	29.3										
		Broad Street		1971	B1		200	4			3		
		High Street		1971	B1		331	4			3		
		New Conduit Street		1971	B1		164	4			3		
		Norfolk Street (part)		1971	B1		114	4					
	NORTH YORKSHIRE												
	SCARBOROUGH	SCARBOROUGH	629.0										4
			97.2										
		Bar Street	40.0	1977	B1		150	1	300	400	1		4,6,13
		Newborough		1977	C	A	80	1	25	15	1		1
		WHITBY	12.0										6
		Baxtergate		1975	B1		260	2			1		4
		Wellington Road		1975	B1		50	2					
	YORK	YORK	104.5										4,6,13
		Castlegate (part)	101.4	1979	B1	A	150	1	250	530	2	bw	5,8
		Coppergate		1979	C	A,T	120	1			1	by	
		Shambles		1979	B1	A	110	2				bz	
		Fossgate		1979	B1	A	180	1				bw	
		Back Swinegate		1979	B1	A	60	1				bw	
		Church Street		1979	B1	A	130	1				bw	

1	2	3	4	5	6	7	8	9	10	11	12	13	14
		(network continued)											
		Colliergate		1979	B1	A	80	1				bw	
		Coney Street (part)		1979	B1	A	90	1				bw	
		Coney Street (part)		1979	B1	A	130	2				bx	
		Davygate		1979	B1	A	140	1				bx	
		Feasegate		1979	B1	A	75	2				bw	
		Goodram Gate		1979	B1	A	170	1				bw	
		Grape Lane		1979	B1	A	70	1				bx	
		High Ousegate		1979	B1	A	130	2				bw	
		High Petergate		1979	B1	A	100	2				bx	
		Kingssquare		1979	B1	A	60	1				bx	
		Little Stonegate		1979	B1	A	85	1				bw	1
		Low Petergate (part)		1979	B1	A		1				bw	
		Low Petergate (part)		1979	B1	A	180	2				bw	
		Market Street		1979	B1	A	120	2				bx	
		Minster Gate		C14th	B	A	30	5				bx	
		New Street		1979	B1	A	90	1				bw	
		Parliament Street (part)		1979	B1	A		1				bw	
		Parliament Street (part)		1979	B1	A	160	2				bx	
		Peter Lane		1979	B1	A		2				bw	
		St. Andrewgate (part)				A	90	1				bw	
		St. Sampsons Sqare		1979	B1	A	80	2				bx	
		Spurriergate		1979	B1	A	170	2				cd	
		Stonegate		1979	B1	A	100	1				bw	
		Swinegate		1979	B1			1				bw	
		Walmgate (part)		1979	B1							bw	
NORTHAMPTONSHIRE	CORBY	CORBY											
		Central Shopping Precinct	468.0		B2							cb	4
		Corporation Street*	52.1										1
		Elizabeth Street*	47.7										1
		George Street*											
	KETTERING	KETTERING											
		Newborough Centre	65.9	1975-78	A2			4/5					4
		Gold Street	41.9	1977	C								6
	NORTHAMPTON	NORTHAMPTON											
		Grosvenor Centre	133.5	1974	A2		310		20	575	2 1 2		
		Gold Street	123.2	1974	B1		50						
		The Drapery		1974	B1		200				3		4,13

131

1	2	3	4	5	6	7	8	9	10	11	12	13	14
NORTHUMBERLAND	WELLINGBOROUGH	WELLINGBOROUGH Enclosed Centre	56.0 36.9		A2								4
	BLYTH VALLEY	BLYTH Bowes Street	280.0 61.0 34.4	1980	B1		140	1	150				4
		Church Street (part)		1979	B1		70	1					4
		Commercial Rd/Church St.		1980	B1		135	4					
		Market Street		1979	B1		100	1					
		Parsons Street		1974	B1		35	4					4
	CRAMLINGTON	CRAMLINGTON Covered Malls	53.1 9.2	1973-78	A2		150	4	90		1		4
	TYNEDALE	HEXHAM Fore Street		1972	B1		85	1	75			*ei*	4
NOTTINGHAMSHIRE	ASHFIELD	BULWELL Commercial Road	973.0 100.9		B1							*cb*	4
		SUTTON-IN-ASHFIELD Idlewells Centre* Low Street*	39.9		A2/B2 B1								4
	BASSETLAW	WORKSOP Netherholme Centre	97.6 35.2		B2							*cc*	4
	BROXTOWE	BEESTON The Square	101.8 63.2		B2			4,5					4
	GEDLING	ARNOLD Front Street	93.5 33.2		B1								4
	MANSFIELD	MANSFIELD Four Seasons Centre* Stockwell Gate* Westgate*	95.4 56.5		A2 B B								4
	NEWARK	NEWARK Arrowcroft Development	98.8 24.3		A2/B2						1		1
	NOTTINGHAM	NOTTINGHAM	299.8 290.8						0/50/100	300	4 5 1 1		4,9,13

1	2	3	4	5	6	7	8	9	10	11	12	13	14
OXFORDSHIRE		Albert Street		1973-74	B1		60	5					
		Broadmarsh Centre		1973-74	A2		200	5					
		Listergate		1973-74	B1		140	3					
		Exchange Walk		1973-74	B1		75	3					
		Clumber Street		1973-74	B1		150						
		Pelham Street		1973-74	B1		135						
		Bridlesmith Gate		1973-74	B1		175						
		St. James Street		1973-74	B1		135	5					
		Victoria Centre		1973-74	A2		400						
		Clinton Street*		1973-74	B1								
	OXFORD	OXFORD	504.0										
		Cornmarket Street	108.6	1973	C	T,S,A	200	2	250	700	1		4,6,13
		Queen Street	95.4	1973	C	S,A	125	2			2		3,7,8
		Westgate Shopping Centre			A2		180	5			2		3,8
	SOUTH OXFORD-SHIRE	WALLINGFORD '1 unit'	133.7 5.9	1972	B1		80	3/4					4
	VALE OF WHITE HORSE	ABINGDON '1 unit'	93.1 17.7	1965	A2/B2			4					
SHROPSHIRE	SHREWSBURY	SHREWSBURY	337.0										
		Riverside Shopping Centre	82.3 53.6	1967	B2		130	4	150	370	1		4
	WREKIN	TELFORD Malinslee Town Centre	97.1	1973	A2			5	0		2		4,13
		Local Centres: Dawley	78.1	1980	B1			2/4					4
		Hadley		1970?	B2			1					
		Madeley			B2			4					
		Oakengates		1977	B2			4					
		Randlay			B2			4					
		Sutton Hill		1969?	B2			4					
		Wellington			B1			2					
		Woodside		1971?	B2			4					

133

1	2	3	4	5	6	7	8	9	10	11	12	13	14
SOMERSET	MENDIP	FROME	387.0 79.2 13.1										
		Kingsway Centre		1975	B2		52	4			2		5
		Westway Centre		1973	B2		155	4					4
		SHEPTON MALLET	5.4										
		Town Street		1974	B1/B2		140	4/5			1		4
		STREET	7.9										1
		Crispin Centre		1979	A2		105	3/4			1		4
	TAUNTON DEANE	TAUNTON	81.9 34.8										
		High Street			B1		72						4
	YEOVIL	CREWKERNE	114.3 4.6										
		Falkland Square		1976	B2		32	4					4,5
		ILMINSTER	3.3										
		Swan Centre		1973?	B2			4					4
		YEOVIL	24.9										
		Glovers Walk*		1965	B1		70	4					4
		Middle Street*		1971			85	1/3/4					
SOUTH YORKSHIRE	BARNSLEY	BARNSLEY	1315.0 226.0 74.2						38	96			
		Albert Street East		1975	B1		70	4			2	cd	4
		Cheapside		1975	C		136	4			3	cd,œ	4
		Market Street		1975	B1		242	2/4			1	cd	3
		May Day Green		1975	C		38	2/4				cd,ce	
		Queens Street		1975	A1		78					cd	
		The Arcade		1975	A1		128	2			2	cd	
		Hanson Street		1979	A1		10	1					
		Regent Street		1979	B1		18	1					1
	DONCASTER	DONCASTER	280.0 80.1						175/475	350			
		Arndale Centre			A2		176	5			4		4
		Baxtergate		1975	B1		104	2			0	cd	
		Frenchgate		1975	B1		125	2/4			2		
		Kingsgate		1966	A2/B2		160	5					
		Princegate (part)			A2/B2		54	5					
		Queensgate			A2/B2		106	5					
		Waterdale Centre			A2/B2		335	5					
		Printing Office Street		1977	B1		36	2			1		

1	2	3	4	5	6	7	8	9	10	11	12	13	14
	ROTHERHAM	ROTHERHAM	243.0 83.6										
		All Saints Square*			B2		55						4
		Howard Street*			B2		80						1
		Walker Place*		1979	B2		60		200	1060			1
	SHEFFIELD	SHEFFIELD	566.0 509.7										
		Exchange Street		1974	B2		78	4	350	440	1		4,13
		Fargate		1971	B1		192	2			2	cd	1
		Norfolk Road		1971	B1		105	1			3		
		The Moor			B1		360	4			2	cd	
		Woodhouse Square		1979	B2		50	4					
STAFFORDSHIRE	NEWCASTLE-UNDER-LYME	NEWCASTLE-UNDER-LYME	963.0 119.7 76.6										
		'1 unit'		1978	C								6
	STOKE-ON-TRENT	HANLEY	265.1										
		'1 unit'			C								6
		STOKE-ON-TRENT												6
		Newcastle Ironmarket	261.8	1977	C	S,A						cf	3
SUFFOLK	FOREST HEATH	MILDENHALL	537.0 39.5 24.8										
		Ped. Shopping Precinct												5
		NEWMARKET	12.4											
		Drapery Row*		1973	B1		18	5					1
		Market Street*		1973	B1		80	3					1
		Rookery Shopping Prcnt.*												
	IPSWICH	IPSWICH	122.8 120.1										
		Museum Street					100				0		6,13
		Tavern Street					320				1		1
		Tower Street					80				0		1
		Westgate Street		1975	C	S,A,U,D,T	300		70	750	2	cg	3,6
		Princes Street					175						1
		Buttermarket					220						1
		Carr Street*					200						1
		Cornhill*												1
		Providence Street*												1

1	2	3	4	5	6	7	8	9	10	11	12	13	14
	MID-SUFFOLK	St. Lawrence Street*	60.4										1
		St. Stephens Lane*	8.5										1
		STOWMARKET	71.2										
		Buttermarket	24.7	1970	B1		30	2					1,5
	ST EDMUNDS BURY	BURY ST EDMUNDS	88.9										
		Buttermarket	7.0	1967	B1		130	0			0	*oh*	5
		Cornhill		1967	B1		210	0			2	*oh*	
		Skinner Street		1967	B1		130	0			0	*oh*	5
		Traverse		1967	B1		50	0			2	*oh*	
	SUFFOLK COASTAL	WOODBRIDGE	91.26										
		New Street	50.6	1977	B1		30	4/5					1
	WAVENEY	LOWESTOFT	981.0										
		London Road North	115.2	1979	C			2/4	70	80	1		5,6
SURREY	ELMBRIDGE	HOLESEY	49.5										
		Hurst Park		1967	B2		40	4					4
		WALTON-ON-THAMES	118.2										
		Precinct adjoining Hepworth Way	54.1	1966	B2	A	150	4					4
	GUILDFORD	GUILDFORD											
		Friary Shopping Centre		1980	A2		80	5	100		1	*oi*	4
		Friary Street		Early70's	B1		450	4/5					4,13
		High Street		Early70's	B1		c.50	2					
		Tunsgate Square		Early70's	B2		c.75	5					
		Phoenix Court*		Early70's	B2			5					
	REIGATE AND BANSTEAD	REDHILL	107.1										
		Station Road	53.5	Late60's	B1		140	2					4
	RUNNYMEDE	ADDLESTONE	75.4										
		Station Road Precinct		Early60's				4					4
	SPELTHORNE	STAINES	96.4										
		Precinct adjacent to bus station	18.4	1979	B2		250	4/5					4
	SURREY HEATH	CAMBERLEY	65.8										4

1	2	3	4	5	6	7	8	9	10	11	12	13	14
	WOKING	Obelisk Street	75.8	1970	B2		500	4					1
		Princes Way	73.1	1970	B2			4					1
	WOKING	Church Street East*	93.2	1976	C		250	4/5					4
		Shopping Precinct*											
TYNE & WEAR	GATESHEAD	BLAYDON	1209.0										
		The Precinct	225.0		B2				0				4
		GATESHEAD	93.2										
		Ellison Street	31.8	1970	B1								4
		Jackson Street			C	A,S,T	170						7,8
		Trinity Square			A2/B2								
		West Street		1970	C		56						
		WREKENTON											
		High Street		1976	C		90						
	NEWCASTLE-UPON-TYNE	NEWCASTLE-UPON-TYNE	308.0										
		Blackett Street	215.1	1974	C	S,A	380		150	125	3		4,7,13
		Clayton Street		1975	B		620				3		3
		Eldon Sq. Shopping Ctre.		1973-81	A2		200				2	1	
		Northumberland Place		1970	B2								
		Northumberland Street		1971	C	S,A	200						
		Saville Row		1971	B		90						3
		Newgate Precinct			A2		150						
		Central Arcade*		Pre1939	A1								
		Handyside Arcade*		Pre1939	A1								
		Princess Square*		1970	B2								
	NORTH TYNESIDE	NORTH SHIELDS	207.0										
		North Shields Shopping Centre	68.1		A2								4
		WALLSEND											
		The Forum		*	A2								4
	SOUTH TYNESIDE	JARROW	177.0										
		Ellison Street	28.6										4
		Grange Road											1
		Monkton Road											1
		The Viking Centre			A2/B2								1
		High Street											1

137

1	2	3	4	5	6	7	8	9	10	11	12	13
		Ormonde Street										1
		Short Row										1
		SOUTH SHIELDS	98.7									
		King Street			B							4
	SUNDERLAND	SUNDERLAND	292.0									
		Blandford Street	213.0		B1		75				2	4,13
		Market Square			B							
		Walworth Way			B							
		WASHINGTON (New Town)	25.2									
		The Galleries			A2		470		0	0	1	4
WARWICKSHIRE	NUNEATON	BEDWORTH	456.0									
		Ped. Precinct	107.5		A2							4
		NUNEATON	40.2									
		Ped. Precinct	66.2		B2							4
	RUGBY	RUGBY	83.6									4
		High Street*	57.6		C							6
		Sheep Street*			C							6
	STRATFORD-ON-AVON	STRATFORD-ON-AVON	94.8									4
		Shopping Precinct	18.4		A2		285	5	270	570	1	4,13
	WARWICK	KENILWORTH	111.5									
		Ped. Precinct	19.9		A2							4
WEST MIDLANDS	BIRMINGHAM	BIRMINGHAM	2790.0						200	150	4	1,4,5,13
		Big Top Arcades (New St/ High St.)	1096.0		A2							
			990.4									
		Bull Street		1974	C		792	1			0	
		Cannon Street (part)		1973	B1	A,S	89	1				
		Cherry Street		1973	A1		95				2	3,8
		City Arcade		Pre1939	B1							
		Corporation Street			B2							
		Corporation Square			A1		160					
		Great Western Arcade		Pre1939	C							
		High Street			B2	A,S	376	1			4	
		Martineau Square			B2		135					3,8
		Martineau Way			B2							

1	2	3	4	5	6	7	8	9	10	11	12	13	14
		(network continued)											
		Minories											
		North Western Arcade		Pre1939	A								
		Old Square			A1								
		South Eastern Arcade		Pre1939	B2								
		Temple Row		1973	A1		125						
		Union Passage			B1		155						
		Union Street		1973	B1		130	1					
		Warwick Passage			B1			1					
		Windsor Arcade		Pre1939	A1								
		Bull Ring Shopping Cntre			A2								
		St. Martins Circus			B2								
		Birmingham (New Street) Shopping Centre			A2								
		Dalton Way			B1								
		Priory Walk			B2								
		Burlington Arcade		Pre1939	A1								
		Piccadilly Arcade		Pre1939	A1								
		Poolway Centre*			B2								4
		CASTLE VALE											
		Castle Vale Centre			B2								4
		EDGBASTON											
		Auchinlock Square			B2								4
		ERDINGTON											
		Central Square			B2								4
		FOX & GOOSE											
		Fox & Goose Centre			B2								4
		KINGSTANDING											
		Circle Shopping Centre			B2								4
		NEWTOWN											
		Newtown Centre			B2								4
		NORTHFIELD											
		Grosvenor Centre			A2								4
		PERRY BARR											
		Linton Square			B2								
		SUTTON COLDFIELD	81.9								5		
		Gracechurch Centre*			B2						2		
		Sainsbury Centre*			A						4		
		WYLDE GREEN			B2								
		Terminus Centre (part)			A						1		
	COVENTRY	BELL GREEN	336.0						200				4,5,9

139

1	2	3	4	5	6	7	8	9	10	11	12	13	14
		Riley Square Precinct											
		CANNON PARK	328.7		B2		160	4	200	600	5 0 1 0		4
		Centre			A2	T,A,S	170	4					
		COVENTRY		1975									4,5,13
		Broadgate		1965-68	C		50	4					3
		Bull Yard		1961	B2		50	4					
		City Arcade		1969-73	A2		100	4					
		Hertford Street		1979-80	A1/B1	A	210	4					
		High Street		1957-60	C		120	4					
		Lower Precinct		1955	B2		150	4					
		Market Way		1951-54	B2		190	4					
		The Precinct		1960	B2		120	4					
		Shelton-Square		1955	B2		60	4					
		Smithford Way		1971	B2		185	4					
		Warwick Row			B1		170	4					
		Cross Cheaping Arcade*	294.0	1961	A		50	1	0				4,5
	DUDLEY		183.6										
		Churchill Precinct		1962-69	B2		210	4/5			3 1 1		4,5
		Fountain Arcade		1926	A1		83	3/4				cj	
		High Street/Market Place		1975	B1		390	2					
		Stone Street		1975	B1		87	2					
		Trident Centre		1973	A2		219	4					
		Birdcage Walk	53.6	1964	B2		95	4	0		4 1		4,5
	HALESOWEN												
		Hagley Street		1973-74	B1		160	4					
		High Street		1967-68	B1		90	4					
		Pelkingham Street		1967-68	B1		75	4					
		The Precinct		1965-66	B2		145	4					
	STOURBRIDGE		18.7						30	30	1 0 1 0		4,5
		Foster Street*		1976	B1		55	2				cj	
		High Street*		1975	B1		390	2				cj	
		Market Street (part)*		1976	B1		70	2				cj	
		Ryemarket*		1973-75	A2		155	4					
	SANDWELL		330.0										
		CRADLEY HEATH									2		4,5
		Market Precinct		1968	B2		165	4					
		SMETHWICK											4,5
		Tollgate Precinct		1976	B2		205	4					
		SWAN VILLAGE							13				4
		Tivoli Centre			B2								
		WEDNESBURY									2 1		4,5
		Camphill Precinct		1967	B2		125	4					

1	2	3	4	5	6	7	8	9	10	11	12	13	14
	SOLIHULL	WEST BROMWICH											
		High Street (part)		1972-74	B1			4			3		4,5
		Sandwell Centre	192.0	1972-74	A2		590	4	0		1		
		CHELMSLEY WOOD											
		Shopping Centre			B2								4
		KNOWLE											
		Precinct(off High Street)	105.9		B2								4
		SOLIHULL											
		Drury Lane (part)*			B2								
		Mill Lane (part)*			B2								
		Sainsbury Centre*			A2								4
	WALSALL		273.0										
		ALDRIDGE											
		Shopping Centre	87.1	1972	B2		340	4					4,5
		BROWNHILLS											
		Ravens Court		1977	B2		75	4					4,5
		DARLASTON											
		High Street		1978	B1		30	4					4,5
		WALSALL	183.0										
		The Arcade		1900	A1		110	1			2		4,5
		Digbeth Square		1965	B2		60	4					
		Old Square		1965	B2		125	4					
		Park Street (part)		1975	B1/C	A,S	165	4			3		3
		Saddlers Centre		1980	A2		190	4			3		
		Townend Square		1965	A		105	4					5
		WILLENHALL											
		Stafford Street (part)		1980	B1		50	4					5,5
	WOLVERHAMPTON		269.0										
		BILSTON											
		Market Way*		1974	A2		120	4					4,5
		Pinfold Street (part)*		1972	B1		13	3					
		Stafford Street (part)*		1972	B1		12	3					
		WOLVERHAMPTON	266.0										
		Bilston Street (part)		1973	B1	S,A	45	1					4,5
		Cheapside		1978	B1	S,A	50	1					
		Lich Gates		1973	B1	S,A	55	1					
		North Street (part)		1975	B1	S,A	134	4					
		Queen Square		1973	B1	S,A	25	1					
		St Peters Square (part)		1979	B		73	3					
		Dudley Street		1973	B1	S,A	130	1					
		Farmers Fold		1965	B1		40	5					
		King Street		1973	B1	S,A	120	1					

1	2	3	4	5	6	7	8	9	10	11	12	13	14
		(network continued)											
		Mander Square		1967	A2		265	5					
		Queen Street (part)		1973	B1	S,A	70	1					
		St. Johns Street		1969	B1		59	5					
		Woolpack Alley		1973	B1	S,A	41	1					
		Wulfrun Centre		1970	A2		125	5					
		Blossoms Fold		1975	B1	S,A	49	1/3					
		Lichfield Passage		1975	B1	S,A	33	1					
		Wadhams Fold		1978	B1		85	5					
		Exchange Street*		1973	B1	S,A	42	1					
		Woolpack Street*		1973	B1	S,A	40	1					
	WEST SUSSEX												
	ARUN		610.0										
		BOGNOR REGIS	105.8										
		High Street	32.8	1902	A1		56	3					5
		LITTLEHAMPTON	17.8										
		High Street (part)		1922	A1		47	3					4
		High Street (part)		1979	B1		130	4					
		High Street (part)		1979	B1		55	1					1
	CHICHESTER	CHICHESTER	91.1						145	145			4,6,6,10
		Crane Street	18.8	1977	B1		87	3/4			2	ej	
		East Street (part)		1977	B1		230	2/4			3	ej,ck	
		North Street (part)		1977	B1		242	2/4			1	ej,ck	3
		South Street		1977	C	S,A	65	1/2/4			2	ej,az	3
		West Street		1977	C	S,A	70	1/2/4				ej,cm	
	CRAWLEY	CRAWLEY	86.2			T			119	204			4,5
		Broadwalk	66.7	1959	A2		110	4					
		The Broadway		1980	C		190	4					
		The Martlets		1959	A2		105	4					
		Queens Square		1974	B1		113	4					
		Queensway		1971	B1		75	4					
	HORSHAM	HORSHAM	85.4										4,5
		Middle Street*	25.6	1950	B1		50	3/4					
		West Street*		1975	B1		250	4					
	MID-SUSSEX	BURGESS HILL	100.2										
		Church Road	19.2	1972	B1/B2		175	4					5
		The Martlets		1972	B1/B2		160	4					4
	WORTHING		88.2							406	2		
											2		
											0		
											1		

1	2	3	4	5	6	7	8	9	10	11	12	13	14
WEST YORKSHIRE		**WORTHING**											
		Montague Street	84.4	1969-74	B1		370	4		850	3 0		4,5
		Warwick Street		1978	B1		116	5			1 0		
	BRADFORD	**BRADFORD**	2053.0										
		Arndale Centre	463.0	1975?	A2		135	5	265	265/163	1 1		4,13
		Darley Street (part)	288.0	1976	B1		124	3/4			1 3		
		Broadway (part)		1971	B1		18	3/4					
		Broadway (part)		1975	B1		59	2					
		Market Street (part)		1977	C		50	2					3
		KEIGHLEY	55.1										
		Cooke Lane*		1970	B2		150	5					4
		Low Street (part)*		1971	B1		75	2					
	CALDERDALE	**HALIFAX**	194.0										
		Cheapside (part)	89.4	1976	B1		60	2				cn	4
		Cornmarket		1976	B1		55	2				cn	
		Russell Street		1976	B1		58	2				cn	
		Southgate		1973	B1		72	2				cn	
		Old Cock Yard		1973	B1		15						
	KIRKLEES	**HUDDERSFIELD**	369.0										
		Macauley Street (part)*	128.6	1974	B		50	0					1,4
		Macauley Street (part)*		1974	B1		180	1					
		New Street*		1972	B1/B2		90	4/5					
		Upperhead Row*		1974	B1			1					
	LEEDS	**LEEDS**	738.0										
		Albion Street (S. part)	481.3	1979	B	T	180	0	300	100	3 3		4,13
		Albion Place		1979	B		135				1		8
		Bond Court		1973	B		205	0					
		Bond Street		1970?	B		115	0					
		Central Road		1970?	B		190	0					
		Commercial Street		1970	B	D,A,U	330	0					
		Headrow		1975	C			0					
		King Charles Street		1975	C			0					
		Kirkgate		1972	B		110	0					
		Lands Lane		1973	B		205						
		Trinity Street		1973									
		Vicar Lane		1973									
		Fish Street		1970?	B			0					

143

1	2	3	4	5	6	7	8	9	10	11	12	13	14
	WAKEFIELD	Merrion Shopping Centre			A2		200	0					
		Queen Victoria Street		1973	B		120						
		Park Square South (pt.)		1977	B								
		Briggate*	289.0	1972									1
		Woodbournehouse	17.1	1979		U						8	4
		OSSETT											
		Bank Street (part)*		1975	B1		104	3/4					1
		Dale Street*		1975	B1		90	3/4					1
		Station Road*		1975	B1		67	3/4					1
		Town End*		1975	B1		185	3/4					1
		Towngate*	30.8		B1								4
		PONTEFRACT											
		Bridge Street*		1976	B1		30	2					1
		Church Lane*		1973	B1		63	4					1
		Gillygate*		1976	B1		65	2					1
		Headlands Road*		1972	B1		93						1
		Middle Row*		1976	B1		100	1					1
		Salter Row*		1973	B1		100	4					
		Shoe Market*		1972	B1		70						
		Woolmarket*		1976	B1		40						
		WAKEFIELD	55.8										4
		Bread Street		1975	B1		85	0					
		Cross Street		1975	B1		75	0					
		Kirkgate (part)		1975	B		200	0					
		Little Westgate		1975	B1		135	0					
		Northgate (part)		1975	B1		80	0					
		Southgate (part)		1975	B1		30	0					
		Teall Street		1975	B1		50	0					
		Springs		1975	C	S,A	200	0					3
		Westmorland Street		1975	C	S,A		0					3
		Union Street		1975	C	U	160	0					
WILTSHIRE	KENNET	DEVIZES	486.0										
		The Brittax	64.4		B?	U	50?						5
			9.8										
	SALISBURY	SALISBURY	100.9						150	1000	2		
		Laybridge Street	33.7				350?				1		1,3
	THAMESDOWN	Neil Road	139.2						100	40			1
		SWINDON	90.0										5
													5,13

144

1	2	3	4	5	6	7	8	9	10	11	12	13	14
		Bridge Street		1974-80	B1		50/90	4				*ep*	1
		Brunel Centre		1973-78	A2		112	4					
		Brunel Centre Surrounds		1973-78	B2		400	4					1
		Canal Walk		1975-77	B2		200	4					1
		College Street		1974-75	B1		15	4					1
		Edgware Road		1974-75	B1		10	4					1
		The Parade		1963	B2		150	1/5					1
		Regent Circus (Theatre Sq)		1970	A2		120	4					1
		Regent Street		1974-75	B1		325	4					4
	WEST WILTSHIRE	Temple Street		1975	B1		40						
			86.7										
		MELKSHAM	9.7										
		a scheme*			A								
		a scheme*			A								
		High St. Link*			B								
		TROWBRIDGE	19.0										
		Fore Street (part)										*aq*	4
WALES													
CLWYD	ALYN & DEESIDE		358.2										
		BUCKLEY	65.4										
		Shopping Precinct	11.9	1971	A2/B2		108	4			2 1		4
	DELYN		56.8										
		MOLD											
		Daniel Owen Shopping Pct	8.2	1975	A2		69	4			2 0 2 1		4
		High Street		1977	B1			2				*aj*	
	RHUDDLAN		47.3										
		PRESTATYN											
		Shopping Precinct	14.2	1972	B2		37	4			1 1 0 1		4
		RHYL											
		Arcades	20.8	1979	A1/A2		92	3			0 2 0 1		4
	WREXHAM MAELOR		104.9										
		WREXHAM	37.6										
		Henblas Street		1975	B1	A	45	2/3				*ek,ar*	5
		Hope Street		1977	B1	A	248	2				*ek,ar*	4
		Regent Street (part)		1977	B1		78	2				*ek,ar*	
		Queen Street		1977	B1	A		2				*ek,ar*	
		Lord Street		1980	B2		125	4					
DYFED	LLANELLI		314.5					1					
		Market Hall & Arcades*	76.8	Pre1939	A1								

1	2	3	4	5	6	7	8	9	10	11	12	13	14
GWENT	MONMOUTH	LLANELLI											
		Stepney Street	26.0	1971	B1		210	4	210		1		5
		Vaughan Street		1971*76	B1		120	4					5
		MONMOUTH	439.9										
		Church Street	64.2	1972	B1		70	1/4	0		1		5
			6.1										
	TORFAEN	CUMBRAN	88.3										
		Town Centre Complex	41.0	1976	A2/B2		c.1250	2			8		5
		PONTYPOOL	36.7										
		Commercial Street (part)		1980	C		50				1		5
GWYNEDD	ARFON	BANGOR	219.9										
		High Street	52.3	1973	B1		170	2/4					4
		Wellfield Centre	12.5	1965-66	A2			5				c8	4
MID-GLAMORGAN	MERTHYR TYDFIL	MERTHYR TYDFIL	530.7										
		High Street (part)*	63.6	1975	B1		225	4					4
		Shopping Precinct*	54.6	1970	B2		140	5					4
	OGWR	BRIDGEND	123.4										
		Adare Street*	14.0	1973	B1		160	2					4
		Shopping Precinct*		1974	A2		100	5					4
	TAFFELY	CAERPHILLY	85.9										
		Pentrebane Street	40.4	1973	B1		100	2				ct	4
		PONTYPRIDD	34.2										
		Shopping Precinct*		1965	A2		100	5				cu	4
		Taff Street*		1978	C	A	500	1				cv	4
POWYS	RADNOR	KNIGHTON	98.8										
		High Street	18.3	1980	B1		101	3			1		4
SOUTH GLAMORGAN	CARDIFF	CARDIFF	389.9	1980					500	65	2		4
			286.4								2		4
			271.5								1		

1	2	3	4	5	6	7	8	9	10	11	12	13	14
		Andrews Arcade		Pre1900	A1		76	3					
		Charles St. (North)		1975–78	B1		46	2					
		Churchill Way (part)		1975–78	B1		30	2					
		Frederick Street		1975–78	B1		115	2					
		Church Street		1975–78	B1		70	2					
		Trinity Street		1975–78	B		130	2					
		Victoria Place		1975–78	B		60	2					
		Duke Street Arcade		Pre1900	A1		25	3					
		High Street Arcade		Pre1900	A1		55						
		Morgan Arcade		Pre1900	A1		55	3					
		Royal Arcade		pre1900	A1		55	3					
		Oxford Arcade		1966	A2		112	3					
		Queen Street		1975–78	B1		510	2					
		Queen Street Arcade		Pre1900	A1		53	3					
		Wobgan Arcade		Pre1900	A1		65	3					
		Wyndham Arcade		Pre1900	A1		37	3					
WEST GLAMORGAN	NEATH	NEATH	371.9										4
			67.3										4
			27.9										
		Green Street		1976	B1	S/P	135	1					
		New Street		1976	B1	S/P	55	1					
		Orchard Street		1976	B1	S/P	105	1					
		Queen Street		1976	B1	S/P	185	1					
		Wind Street		1976	B1	S/P	70	1					
	SWANSEA	SWANSEA	189.0										
		College Street	168.3	1967	B1	A	200	3	0	550	3		4,13
		Oxford Street		1976	B1/C		640	4					
		Whitewalls		1976	C	A,S,T		2					
SCOTLAND													
CENTRAL	FALKIRK	FALKIRK	263.3										
		Callender Rigs	141.0		B2		120	5	0				4
		STENHOUSEMUIR											
		Shopping Centre		1972	B2		300	4					
		GRANGEMOUTH											
		Charlotte Dundas Shopping Centre			B2		90	4					4

147

1	2	3	4	5	6	7	8	9	10	11	12	13	14
	STIRLING	La Porte Precinct	76.1	1968	B2		225	3/4	10	110			4
	STIRLING	Arcade		Pre1939	A1		120	3/4					4
		Thistle Centre		1977	A2		165	4/5					4
FIFE	KIRKCALDY	GLENROTHES Kingdom Centre	328.0 / 146.0	1976	A2		400	4					4
		KIRKCALDY Shopping Arcades	35.4	1973	A		125	4					4
GRAMPIAN	CITY OF ABERDEEN	ABERDEEN High Street	437.2 / 208.4	1973	C	S	430	0	275	350	2		5,13
		St Nicholas Street/ George Street(pt.)		1977	C	S,D	320	0					
HIGHLAND	INVERNESS	INVERNESS Academy Street Arcade	175.4 / 49.4	Pre1939	A1		59	3	75	50	1		4
		Queensgate Arcade		Pre1939	A1		47	3					
		Lombard Street		1965	B	S	43	3					
LOTHIAN	CITY OF EDINBURGH	EDINBURGH Rose Street	742.3 / 472.2		B1	S	425	2	30	335	0		4,13
		St. James Centre			A2		175	5			1		
	WEST LOTHIAN	LEITH Kirkgate Centre	112.3		B2			4			0		4
		LIVINGSTON Livingston Centre	35.0	1976	A2						0		4
STRATHCLYDE	CITY OF GLASGOW	GLASGOW Buchanan Street	2578.0 / 983.5	1972	B		495		100	175	1,1,1,2	e1,cw	13,12

148

1	2	3	4	5	6	7	8	9	10	11	12	13	14
TAYSIDE		Souchiehall Street	396.8	1972	B	D,U	400	4					12
	ANGUS	ARBROATH											
		High Street	84.5	1973	B1		200	2/4					4
	CITY OF DUNDEE	DUNDEE	197.5										
		Murraygate		1976	B1	S	180	2					4
		Overgate Shopping Centre		1962–68	A2/B2		346	4			1	cx	
		Wellgate Centre		1978	A2		220	5	130	280	2		4,5
	PERTH & KINROSS	PERTH	114.7										
		St John Square		1960–65	B2		80x60	4	700	750	0		4,13
WESTERN ISLES		STORNOWAY	30.6										
		Cromwell Street (S.)		1968	B1		50	4/5			0		4
NORTHERN IRELAND													
NORTHERN IRELAND	BELFAST CITY COUNCIL	BELFAST	548.0										
		Ann Street (part)		1976	B	S	200)	2	700)		7	cy	4
		Ann Street (part)		1978	B	S	)	2	)		5	cy	4
		Arthur Place		1979	B	S	40	2	620			cy	
		Arthur Square (part)		1976	B	S	15)	2	670)			cy	
		Arthur Square (part) (S)		1978	B	S	)	2	)			cy	
		Arthur Street		1979	A/B	S	60	2	600			cy	
		Castle Arcade		1970	B	S	80	2	600			cy	
		Castle Lane (part)		1975	B	S	140)	2	560			cy	
		Castle Lane (part)		1978	C	P,S	)	1	)			cz	
		Castle Place		N/A	A2		160	4	620		7		
		Central Arcade		1978–79	B	S	30	2	670			cy	
		Church Lane		1978	B	S	110	2	840			cy	
		College Street (part)		1978	B	S	45	2	390			cy	
		Cornmarket		1976	C	P,S	80	2	680		4	cz	
		Donegal Square West		N/A	B	S	130	1/4	320				
		Fountain Lane		1978	B	P,S	70	2	480				
		Fountain Street		1978	B	S	160	2	370				
		Royal Avenue		N/A	C	P	250	1	620			da,cz	
		Upper Queen Street		N/A	C	P	130	1	200			cz	

149

1	2	3	4	5	6	7	8	9	10	11	12	13	14
		(network continued)											
		Wellington Street		1978	B	S	85	2	240			cy	4
		William Street South		1978	B	S	80	2	700			cy	4
		North Street Arcade		1930s	A1		85	3	820				
		Queens Arcade		1900s	A1		75	3	440				
CHANNEL ISLANDS													
GUERNSEY		ST PETER PORT	52.7										
		Commercial Arcade		1980	A2		190	3	120			db,do	4
		High Street		1970	B1		270	2	135			db,do	4
		La Plaiderie		1970	B1		60	2	380			db,do	
		The Pollet		1970	B1		40	2	370			db,do	
JERSEY		ST HELIER	73.0										
		Bath Road (part)		1979	B1		50	3	360		1	dd	4
		Bath Road (part)		1980	B1		110	2	390		2	dd	4
		Halkett Street		1972	B1		150	2	330		0		
		Market Street		1970	B		40	3	370		1		
		Queen Street		1979	B1		120	3	280				
		Snow Hill		1979	B1		30	3	360				
		Don Street (part)		1970	B1		40	3	240				
		King Street		1970	B1		330	3/4	210				
		Pitt Street		1970	B1		50	2	270				
		Royal Square			B		80	5	170				

NOTES TO TABLE

a *No conversion of pedestrianised street involved*
b *Pedestrianised Saturdays only 10.00-17.00*
c *Access to pedestrian area 10.00-16.00 only*
d *Monday-Friday (10.30-16.30) Vehicles allowed only for loading/*
    unloading goods (except in St Werburgh Street)
    Monday-Saturday (08.00-10.30 and 16.30-18.00) Vehicles allowed
    only for access
    Saturday (10.30-16.30) No vehicles allowed (except buses in
    St Werburgh Street)
    Other times: no restrictions on access
    Disabled drivers may not enter area between 10.30 and 16.30 on
    Saturdays
    Coaches will be allowed into these areas, to the north and east
    of the Cross, except between 10.30 and 16.30 on Saturdays
e *Monday-Friday (08.00-18.00) Vehicles allowed only for access*
    Saturday (08.00-10.30 and 16.30-18.00) Vehicles allowed only
    for access
    Saturday (10.30-16.30) Commercial vehicles over 3 tons unladen
    weight and coaches not allowed; other vehicles allowed only
    for access
    Other times: no restrictions on access
    Coaches will be allowed into areas to north and east of the Cross
    except between 10.30 and 16.30 on Saturdays
f *Monday-Saturday 08.00-18.00 vehicles allowed only for access. No*
    restrictions at other times
g *All traffic prohibited, except buses through Town Hall Square*
h *'Allowed only for access' means that vehicles may enter the area*
    ONLY for setting down/picking up passengers or loading/unloading
    goods
i *Summer only pedestrianisation*
j *Pedestrianisation between 10.00 and 17.00*
k *East part no entry 10.00-17.00 except buses and for access*
l *West part no entry except buses and service vehicles*
m *Pedestrianised weekdays 10.00-16.00*
n *Access only during shopping hours*
o *Redevelopment in Conservation Area*
p *Pedestrians and buses (where necessary) and private vehicles*
    (access only) all day
q *Pedestrians and buses only*
r *Pedestrians and buses only 10.00-18.00 and service access 18.00-10.00*
s *Pedestrians only*
t *Pedestrians only 10.00-16.00 and service access 16.00-10.00*
u *Servicing: there are no properties fronting the length of street*
    pedestrianised
v *Restrictions 23.00-10.00 except Sunday*
w *Half road pedestrianised, half traffic*
x *Closed Saturday only*

y   Private vehicles banned Monday-Saturday 09.30-17.30
z   Loading permitted before 09.30 and after 18.00
aa   Forms part of Lion Walk precinct
ab   Underground servicing
ac   Still under construction
ad   LT objection to loss of access for buses overruled by GLC on
     traffic grounds, supported by Police
ae   Closed to traffic 10.00-19.00 Fridays and Saturdays only
af   Pedestrian area originally uncovered
ag   'Time-sharing' service. Traffic allowed in and out at shopping
     hours
ah   Vehicles allowed in southern end at certain times
ai   Closed Saturdays only, buses diverted. Open to traffic for the
     rest of the week
aj   Pedestrianised only Thursday, Friday, Saturday
ak   Restrictions 24 hours a day seven days a week. Buses only
al   Includes street market
am   Under Charing Cross BR station
an   Private, not maintained by the Council
ao   Part of St Christophers Place pedestrianisation scheme
ap   Vehicles are not permitted to enter between 11.00 and 20.00
aq   Continues with the Arches
ar   Part of Carnaby Street pedestrianisation
as   Continues into St Christophers Place
at   Access (service vehicles only) between 07.00 and 19.00
au   Newport Street, Oxford Street, Victoria Square combine to form
     one large precinct
av   Pedestrianised area is total length for seven streets
aw   Pedestrianised area part covered by canopy
ax   Servicing mostly below street level
ay   Part-pedestrianised
az   Servicing anytime
ba   Closed for access Saturdays 10.00-18.00
bb   Pedestrianised Wednesday and Saturday only
bc   Buses at all times - goods vehicles allowed access at all times
bd   Access at all times Sunday, and 17.30-11.00 Monday-Saturday
be   Closed to traffic Saturdays 10.00-17.00
bf   Chancery Walk, Fleet Walk, Howe Walk, The Mall, Market Square -
     all part of central area redevelopment phased 1967-1972
bg   Servicing not required between Harris and Jackson Street
bh   Judges entrance to Courts, no other premises serviced
bi   No access to traffic or loading/offloading of service vehicles
     between the hours of 10.30 and 16.00
bj   Access is permitted only between 11.00 and 17.00 and loading/un-
     loading. Servicing forbidden in peak hours
bk   March 1980 for six months
bl   Servicing permitted between 10.00 and 16.00
bm   Trolley servicing from Partner Street

bn   All shops on Houghton Street face onto other streets
bo   Only southern end of street is paved
bp   Loading Monday-Saturday 08.00-09.30 and 16.30-18.00
bq   Servicing temporarily from road. Will later change to the rear
br   Unadopted
bs   Seasonal
bt   Dove Street only partially closed to allow for delivery to
     local brewery
bu   Pedestrianised Saturdays only - may become Monday to Saturday
bv   Pedestrianised Saturdays only
bw   York - Back Swinegate Front servicing limited 08.00-18.00 Monday-
     Saturday (generally other vehicles prohibited, with exceptions
     for access to premises off street)
bx   Coney Street part, frontage restricted 08.00-18.00 Monday-Saturday
     but within that time no servicing (and no other vehicles) per-
     mitted between 11.30 and 16.30
by   Coppergate Front servicing limited 08.00-18.00 Monday-Saturday
     (generally other vehicles prohibited, except buses)
bz   Shambles 'Access only' street for full 24 hours, with no servicing
     (and no other vehicles) permitted between 10.30 and 15.30
ca   Stonegate No vehicles between 10.30 and 05.00 the next day.
     Servicing permitted 05.00-10.30
cb   Incomplete because charge would be made reflecting staff time
     and administrative costs
cc   Pedestrian centre partly covered
cd   Pedestrianised 10.00-18.00
ce   Buses experimentally excluded until March 1981
cf   Road width narrowed to give partial pedestrianisation
cg   Partial pedestrianisation 09.30-16.30 Monday-Saturday
ch   Closed to traffic only on market days (Wednesday and Saturday)
     No servicing possible from these streets when the pedestrian
     scheme is in force
ci   Closed to vehicles on Saturdays
cj   Pedestrianised Saturdays only
ck   Night servicing permitted to shop fronts 18.00-09.00
cl   West side of street cannot be rear serviced, hence front servicing
     both during day and night and night servicing as ck
cm   Cellars to public house are at front and drays must service from
     there
cn   Access for servicing allowed between 16.00 and 10.30. At other
     times vehicular traffic is prohibited
co   Plaza scheme
cp   Bridge Street closed Saturdays 10.00-18.00
cq   Saturdays only - limited access
cr   Limited access
cs   Additional closure 11.00-16.30 Monday-Saturday
ct   Prohibition of driving 10.00-17.00 Saturdays
cu   Prohibition of driving from 10.00-16.00 on Saturdays
cv   Limited to buses and access 09.00-17.30 Monday-Saturday

153

cw Service access permitted 16.00-11.00.   Total ban between 11.00-
        16.00
cx Service vehicles are allowed access to the street between 11.00
        and 16.00
cy Access for delivery vehicles only 18.00-11.00
cz Only permit holders can bring a car inside security barriers for
        servicing purposes
da Security gates stay open to pedestrians after 18.00
db Servicing: all unloading of goods vehicles must be completed by
        10.00
dc No buses are allowed and between 10.00-18.00 other vehicles are
        allowed only with a Police permit
dd Pedestrianised 10.00-17.00

FUTURE PLANS

ea Chester City Council propose to ban buses from The Cross and
        Northgate in the near future
eb Major shopping development proposed
ec Bus station under construction
ed Further stages under construction
ee Further stages planned
ef Further stages being implemented
eg Phase Two - High Street
eh 610m proposed pedestrianisation in stages.   Total length will
        eventually be 930 m.   There are rear servicing plans
ei Malls to be extended to 280 m in next phase of development, 1982
ej Paving will be completed 1981
ek Will be purpose built pedestrian street in 1980
el Ban of vehicles may be extended from 11.00-17.00 or 18.00

County/District/Place	Frequency Bus stops, direct distance in m			
	0-50	51-100	101-150	151-200
Avon/Bristol/Bristol	21	15	13	15
Bedford/Bedford/Bedford	13	6	1	1
/Luton/Luton	10		2	
Cheshire/Chester/Chester	17	8	4	
Cleveland/Middlesbrough/Middlesbrough	5	3	3	2
Durham/Durham/Durham	7	6		12
/Easington/Peterlee			1	1
/Seaham	2	2		1
/Sedgefield/Newton Aycliffe	2	1		
/Wear Valley/Crook	3	1		
Greater London/Barking		4	1	1
/Redbridge	1	2	1	1
/Wandsworth	8	3	2	2
Greater Manchester/Bolton/Bolton	3	2		
/Bury/Bury	2	1		2
/Manchester/Cheetham Hill	1		1	
/Chorlton	2		2	
/M'chester	11	2	2	3
/Sale	1	3	1	
/Oldham/Oldham	2	2		1
/Rochdale/Middleton	1		2	
/Rochdale	3	1		
/Salford/Eccles	2	2		
/Stockport/Stockport	7	2		
/Tameside/Hyde	4	1		
/Trafford/Altrincham	3			
/Wigan/Wigan	1	1	6	2
Hampshire/Basingstoke/Basingstoke	1	1	1	
Hertfordshire/Watford/Watford	5	7	5	3
Kent/Ashford	4		1	2
/Canterbury	4		1	2
Lancashire/Blackpool	4	2	2	
/Burnley	1	2		1
/Rossendale/Rawtenstall		1		2
/Wyre/Poulton-le-Fylde	4	2		
Merseyside/Liverpool	18	2	7	10
/Sefton/Southport	6	6	2	3
Northumberland/Blyth Valley/Blyth	1	1		2
Nottingham/Nottingham	30	21	10	3
South Yorks/Barnsley	4	5	3	
/Doncaster	1		1	1
/Rotherham	2	2	2	3
Suffolk/Suffolk Coastal/Woodbridge	1	1	3	

County/District/Place/Street	Frequency Bus stops, direct distance, m			
	0-50	51-100	101-150	151-200
West Midlands/Birmingham	48	8	24	22
/Sandwell/Smethwick	4	1	2	4
/Wednesbury				1
/Wolverhampton	26	22	4	3
Wiltshire/Thamesdown/Swindon	8	5	5	5
Highland/Inverness	3	4		
Dundee	4	2	3	1
St Helier/Bath Road part	2			2
/Bath Road part		2		
/Don Street part		1	1	1
/Halkett Street	1	1		
/King Street	1		2	
/Market Street			2	1
/Pitt Street		1	1	
/Queen Street	2			3
/Royal Square			2	2
/Snow Hill	2		2	
Totals (this sample)	313	168	128	131

11
references

Anderson, Alastair (1981)   Just another human being   *World Health* January

Appleyard, D and M Lintell (1970) *Environmental Quality of City Streets* Berkeley   Institute of Urban and Regional Development

- (1975)   Streets dead or alive?   *New Society*   3 July   pp 9-11

Arrive (1971)   Article on Hereford

Association of District Councils (1978)   *Bus operation and traffic management*

Automobile Association (1979a)   *Town Plans*   Second Edition

- (1979b)   *Around Britain's seaside*

Ball, R and R Brooks (1976)   Planning and road design for bus services   *Chartered Municipal Engineer*   Vol 103   September

Barnsley MBC (1975)   *Barnsley Town Centre Local Plan Report of Survey - Transportation supplement*

- (1977)   *Draft Barnsley Town Centre Local Plan*

Bendixson, Terence (1977)   *Instead of Cars* Harmondsworth   Penguin Books

Bennison, DJ and RL Davies (1977)   The local effects of city centre shopping schemes: a case study   *PTRC summer annual meeting University of Warwick 27-30 June*

Betts, I (1975)   Queensmead given over to pedestrians   *Surveyor* 19-26 December

Bishop, D (1975)   User response to a foot street *Town Planning Review* Vol 46   January

Boeminghaus, Dieter (1979?)   *Pedestrian areas*   Stuttgart   Karl Krämer Verlag

Borough of Thamesdown (1980)   *Personal communication*

Bouchier, David (1981)   You must go to the mall *New Society* 26 February

Bradburn, PD and DI Hurdle (1981) *Planning for buses in Greater London: a guide to infrastructure requirements for buses* London   Greater London Council

Brambilla, Roberto and Gianni Longo (1976a)   *A handbook for pedestrian action* Washington   US Government Printing Office   Footnotes No 1

- (1976b) *The rediscovery of the pedestrian: 12 European Cities* Washington US Government Printing Office Footnotes No 2

- (1976c) *Banning the car downtown - selected American cities* Footnotes No 3

- (1976d) *For pedestrians only: planning design and management of traffic-free zones* London Architectural Press

Brambilla, Roberto Gianni Longo and Virginia Dzurinko (1977) *American urban malls - a compendium* Washington US Government Printing Office Footnotes No 4

British Multiple Retailers Association (1980) *Guidelines for shopping (draft)* London

Building Design (1981) *Article on Thamesmead shopping centre* March

Central Office of Information (1980) *Reference Pamphlet No 1, local government in Britain* HMSO

Central Statistical Office (1978) *Social Trends No 9* London HMSO

- (1980) *Social Trends No 11* London HMSO

City of Birmingham (n.d.) *City Centre pedestrianisation scheme synopsis*

City of Dundee (1976) *Pedestrianisation of Murraygate - survey information*

- (n.d.) *Central Area Local Plan - survey report*

City of Lincoln (1980) *Copy of personal communication from Director of Planning & Architecture to a County Planning Officer*

City of Newcastle upon Tyne (1979) *City Centre Local Plan public participation report* City Planning Department July

City of Oxford Motor Services Ltd (1978) *A report examining the use of streets in Oxford Central Area by buses*

City of Portsmouth (1976) *Palmerston Road Shopping Centre*

Civic Trust (1976) *The effects of pedestrianisation on trade: statistics* and *List of pedestrianised schemes*

Clyde, CA (1976) *The pedestrianisation of Church Street Liverpool: a survey of users' attitudes* Working Paper 52 Leeds Institute for Transport Studies

Cooper, JSL PN Daly and PG Headicar (1979) Accessibility analysis *Traffic Engineering & Control* Vol 20 January

Copley, G and MJ Maher (n.d.) *Pedestrian movements: a review* Leeds Institute for Transport Studies

Cullen, Gordon (1971) *Townscape* London Architectural Press

Dalby, E (1973) *Pedestrians and shopping centre layout: a review of the current situation* Transport & Road Research Laboratory Report 577 Crowthorne

- (1976) *Space-sharing by pedestrians and vehicles* TRRL Report LR 743 Crowthorne

Dalby, E and AE Williamson (1977) *Pedestrian and traffic management techniques in Delft: report of a visit made in December 1975* TRRL Report SR 257 Crowthorne

Daor, E and PB Goodwin (1976) *Variations in the importance of walking as a mode of transport* GLC Research Memorandum 487 London

Darlington Borough Council (1980a) *Central Area Plan report of survey: traffic management and pedestrianisation*

- (1980b) *Central Area Plan report of survey: car parking*

Davis, Ives Associates Ltd (1977) Attitudes of traders and the public to the Wakefield pedestrian precinct *Wakefield Express* January 7

Department of the Environment (1975) *National Travel Survey 1972-3* London HMSO

Department of Transport (1978) *Notes on the preparation of pedestrianisation schemes* Local Transport Note 2/78 London The Department

- (1979) *National Travel Survey: 1975/6 report* London HMSO

- (1980a) *Lists of pedestrianised streets* where Orders were made under Section 212 of the Town & Country Planning Act 1971 and Section 1 of the Road Traffic Regulation Act 1967

- (1980b) *Transport statistics Great Britain 1969-79* London HMSO

Design Council (1976) *Street scene* London The Council

Design Council and Royal Town Planning Institute (1979) *Streets ahead* London The Council

Dorset County Council (1978) *Transport Policies and Programmes 1979/80 submission*

Durham County Council (1973) *Experimental closure of Church Street Seaham. Report of County Planning Officer*

- (1974) *Crook Town Centre – Hope Street pedestrianisation*

- (1975) *Stanley Town Centre – opinion surveys*

Edminster, Richard and David Koffman (1979) *Streets for pedestrians and transit: an evaluation of three transit malls in the United States* Washington Department of Transportation UMTA

Elkington, John Roger McGlynn and John Roberts (1976) *The pedestrian: planning and research* London Transport & Environment Studies

Fowler, Norman (1980) Current policy *(Opening address) Conference on walking* London Policy Studies Institute

Fruin, John J (1970) *Designing for pedestrians: a level of service concept* PhD dissertation, Polytechnic Institute of Brooklyn

- (1972) Pedestrian vs highways: the pedestrian's right to urban space *Highway Research Record* No 403 pp 28-36

Ganguli, BK (1974) *Pedestrian delay studies* Research Memorandum 439 London Greater London Council

Garbrecht, Dietrich (1976) Pedestrian factors and considerations in the design or rebuilding of town centres and suburbs: Paper to *International Conference on Pedestrian Safety* Haifa 20-23 December

Garton, Penelope M (1977) *Structured attitude surveys in five shopping streets before and after implementation of traffic schemes in Central Barnsley* Working Paper 101 Leeds Institute for Transport Studies

Gateshead Borough Council (n.d.) *Gateshead Bus Demonstration Project*

George Godwin Ltd (1974) *Directory of Official Architecture & Planning 1974-75*

- (1980) *do. 1980-81*

Glasspoole, C (1980) Article on Central Manchester *Building Design* 12 December

Goddard, GDS R Tripp and AP Young (1977) *Buses and Pedestrian Areas* Glasgow Greater Glasgow Passenger Transport Executive

Gray, John E (1965) *Pedestrianised shopping streets in Europe* Edinburgh Pedestrians Association for Road Safety

Greater London Council (1972) *GLC Study Tour of Europe and America: pedestrianised streets* London The Council

- (1974) *Car ban proposed for 24 London shopping streets* Press Release 228 April 30

- (1975) *The results of an attitude survey taken during the temporary closure of Heath Street Hampstead in November 1974*

- (1976) *Greater London Development Plan : Notice of Approval*

- (1977) *Pedestrianisation schemes - progress report at December*

- (1978) *Leicester Square pedestrianisation Stage V* Report CAP 291 to Central Area Planning Committee

- (1979a) *Buses in town centres* Report PC358 to Planning & Communications Policy Committee 12 January

- (1979b) *Transport Policies and Programme 1980-85*

- (1980a) *Biggest ever shopping development for London*

- (1980b) *It's all systems go for the new Piccadilly* Press Release June 9

Greater Manchester Council (1976) *Pedestrian survey (of larger town centres)*

- (1981) *Manchester City Centre - pedestrianisation Stage II*

Gwynedd County Council (1980) *Personal communication*

Hillman, Mayer and Anne Whalley (1979) *Walking is transport* London Policy Studies Institute

Hillman, M I Henderson and A Whalley (1973) *Personal mobility and transport policy* PEP Broadsheet 542

Humberside County Council (n.d.) *Summary of pedestrian and trader interviews*

Institute of Traffic Engineers (1966) *Traffic planning and other considerations for pedestrian malls*

Jennings, A CH Sharp and D Whibley (1972) *Delivering the goods: a study of the Watford service-only precinct* London Freight Transport Association

Jones, SR (1981) *Accessibility measures: a literature review* TRRL Laboratory Report 967 Crowthorne

Jonquiere, Peter (1981) Pedestrian priority - the 'Woonerf' principle *Town & Country Planning Summer School Nottingham*

Kerridge, MSP (1979) *Bus priority and marketing schemes* London Confederation of British Road Passenger Transport

Lee, Michael and Elizabeth Kent (1975) *Caerphilly Hypermarket Study Year Two* London Donaldsons

- (1977) *Brent Cross Study* London Donaldsons

Llanelli Borough Council (1980) *Personal communication*

London Amenity and Transport Association (1973) *Pedestrians in London - the need for a policy*

London Borough of Enfield (1980) *Personal communication*

London Borough of Hammersmith & Fulham (1980) *King Street pedestrianisation* Planning Policy Committee 9 April

London Borough of Lewisham (1977) *Catford Local Plan - Shopping topic*

London Transport (1974) *Buses in pedestrian areas* Unpublished report

Lund, Kenneth (1974) *Pedestrian ways* London Borough of Newham

Lynch, Kevin (1960) *The image of the city* Cambridge MIT Press

Mid Glamorgan County Council (1980) *Personal communication*

Mitchell, CGB and SW Town (1977) *Accessibility of various social groups to different activities* TRRL Supplementary Report 258 Crowthorne

Monheim, Rolf (1975) *Fussgängerbereich* Köln Deutscher Städtetag

Morrell, James (1980) Growth in the next decade *Food Manufacturers'*
Federation Conference Grosvenor House London 17 April

Multiple Shops Federation (1963) *The planning of shopping centres*

Municipal Journal (1975) *Article on Durham* 10 October

Murata, Takahiro (1978) Creation of pedestrian streets in city centres
<u>in</u> US Dept of Transportation *Road user information needs, pedestrian*
movement and bicycle travel patterns Publication HS-026 724 (TRR-683)

Murray, RA and PD Ennor (n.d.) *The readmission of buses to a pedes-*
trian street - a Bus Demonstration Project in Gateshead Report
CNS23 Newcastle Transport Operations Research Group

Myatt, PR (1975) *Carnaby Street study* Research Memorandum 466 London
Greater London Council

National Bus Company (1975a) *Bus Priority Schemes* Research Report 7

- (1975b) *Canterbury City pedestrianisation scheme - passenger*
reaction to pedestrianisation and the consequent transfer of bus
stops to Gravel Walk Report MRO35

- (1978) *Bus priority schemes* Report 19

National League of Cities (1980) *Transportation and the urban envir-*
onment Washington US Department of Transportation

Norfolk County Council (1974) *Norwich: London Street - an appraisal*
of costs and benefits

- (1975) *Pedestrianisation in King's Lynn*

North Atlantic Treaty Organisation (1976) *Bus priority schemes* Crowthorne Transport & Road Research Laboratory

OECD's Special Research Group on Pedestrian Safety (1977) *Chairman's*
report and reports of three sub-groups Drowthorne Transport &
Road Research Laboratory

Oeding, Detlef (1963) Verkehrsbelastung und Dimensionierung von Geh-
wegen und anderen Anlagen des Fussgängerverkehrs (Traffic loads and
dimensions of walkways and other pedestrian circulation facilities)
<u>in</u> *Strassenbau und Strassenverkehrstechnik* No 22

Official Architecture and Planning (1970) *Planning for pedestrians*
Special issue June

Older, SJ (1968) Movement of pedestrians on footways in shopping
streets *Traffic Engineering and Control* August

Open University (1975) *Air Pollution, Units 13-15* for course Environmental Control and Public Health

Opinion Research Centre (1970) *Attitudes toward the Saturday pedes-*
trianisation of Deptford High Street Unpublished report to London
Transport

Organisation for Economic Cooperation and Development (1974) *Streets for people* Paris The Organisation

- (1975) *Results of a survey on traffic limitation policies in 300 OECD cities*

- (1977) *Transport requirements for urban communities: planning for personal travel*

- (1978) *Results of questionnaire survey in pedestrian zones* ENV/UT/78.3 5 April

- (1980) *Evaluation of urban parking systems* Paris

Ove Arup Partnership (1973) *Commercial Road Precinct, Portsmouth: monitoring study*

Papanek, Victor (1974) *Design for the real world* London Paladin

Parker, J and J Eburah (1973) Oxford Street experiment *GLC Intelligence Unit Quarterly Bulletin* No 25 December

Parker, John and Charles Hoile (1975) Central London's Pedestrian Streets and Ways *Greater London Intelligence Quarterly* No 33 December

Perkin, George (n.d.) *Streets for pedestrians* Europa Nostra

Plowden, Stephen (1980) *Taming traffic* London Andre Deutsch

Policy Studies Institute (1980) *Conference on walking* London

Potter, Robert B (1980) What is convenient shopping? *Town & Country Planning* April

Potter, Stephen (1981) The Great Car Robbery *Transport Retort* January

Pushkarev, Boris and Jeffrey M Zupan (1975) *Urban space for pedestrians* Regional Plan Association Cambridge MIT Press

Ramsay, A (1977) *Scope and criteria for pedestrianisation* MA Thesis University of Manchester Dept. of Town & Country Planning

- (1980) *Planning for pedestrians* Edinburgh Capital Planning Information

Richards, Brian (1976) *Moving in cities* London Studio Vista

Roberts, John (1980) Walking for the hell of it *Vole* Vol 3 May 8

Roberts, Margaret and John (1970) Shopping and goods distribution: a new approach *Official Architecture and Planning* January

Robertson, Stuart (1980) *Personal communication*

Rudofsky, Bernard (1969) *Streets for people: a primer for Americans* New York Doubleday

Sainsbury, RB and R Caswell (1977) *Air pollution and pedestrianisation: studies at the Upper Norwood Triangle - Westow Hill traffic experiment* RM496 London Greater London Council

Schiller, Russell (1981) Superstore impact *The Planner* Vol 67 No 2 March-April

Scott, ARN (1979) *Durham City floorscape scheme* City Council

Sheppard, D and Shelagh D Valentine (1980) *Elderly pedestrians: assessments of the relative value of instruction on alternative safety topics* TRRL Supplementary Report 601 Crowthorne

Smith, BA (1977) Central Area development, the German style *Town & Country Planning* July-August

South Glamorgan County Council (1980) *Personal communication*

South Pembrokeshire District Council (1980) *Personal communication*

South Yorkshire County Council (1980) *South Yorkshire statistics 1980*

Stewart, JR, B Goodey and AS Travis (1979) *User response to pedestrianised shopping streets* Centre for Urban and Regional Studies RM73

Sudjic, Deyan (1980) Quality street on the rack *Sunday Times* April 27

Surveyor (1977) *Narrow streets to broad ways: all happy in Cardiff*

Thiebault, RW EJ Kaiser, EW Butler and J McAllister (1973) Accessibility satisfaction, income and residential mobility *Traffic Quarterly* 27 pp 289-306

Thomas, Ray and Stephen Potter (1977) Landscape with pedestrian figures *Built Environment Quarterly* December

Todd, JE and A Walker (1980) *People as pedestrians* OPCS London HMSO

Town & Country Planning (1980) *Going places with energy* July/August

Traffic Engineering & Control (1979) *News item* Vol 20 October

Transport & Environment Studies (TEST) (1974) *Pedestrian movement in Brockley* A report for the Transport & Road Research Laboratory

- (1975) *Environmental effects of traffic restraint in Singapore* A report for the World Bank

- (1976) *Improving the pedestrian's environment* Unpublished report to the Department of the Environment

- (1979) *Bus routeing and pedestrianisation in Kingston* Unpublished study for London Transport

- (1980) *Sutton High Street - a study of pedestrianisation* London Transport

- (1981) *Buses and pedestrian areas* London Transport

Van Cort, H Matthys (1978) *The Commons, Ithaca, New York: Bringing people back to the city* Ithaca Department of Planning & Development

Vickerman, RW (1974) Accessibility, attraction and potential: a review of some concepts and their use in determining mobility *Environment and Planning* 6 pp 675-91

Walter, JA (1981)   Family car   *Town & Country Planning*   February
 pp 56-59

Wandsworth, London Borough (1977)   *St John's Road experiment: analysis of questionnaire survey and correspondence*

  - (1979)   *A new look for St John's Road: consultation results*
November

West Midlands Passenger Transport Executive (1979)   *Mode of travel / shopping turnover survey*

West Sussex County Council (1976)   *Pedestrianisation in Chichester*

Whyte, William H (1980)   *The social life of small urban spaces*
Washington   The Conservation Federation

Wood, AA (1966)   *Foot streets in four cities*   Norwich   City Hall

  - (1969)   *Norwich: The creation of a foot street*

  - (1977)   Foot streets and public transport in Roy Cresswell (ed)
Passenger transport and the environment   London   Leonard Hill